Open Source Biomedical Engineering

Hugo Plácido da Silva • Patrícia Justo Bota •
Ana Sofia Cacais do Carmo
Editors

Open Source Biomedical Engineering

Bridging the Gap Between Sensing, Processing, and Visualization

Springer

Editors
Hugo Plácido da Silva
Instituto Superior Técnico
Instituto de Telecomunicações
Lisbon, Portugal

Patrícia Justo Bota
Instituto Superior Técnico
Instituto de Telecomunicações
Lisbon, Portugal

Ana Sofia Cacais do Carmo
Instituto Superior Técnico
Instituto de Telecomunicações
Lisbon, Portugal

ISBN 978-3-032-03654-4 ISBN 978-3-032-03655-1 (eBook)
https://doi.org/10.1007/978-3-032-03655-1

This Springer imprint is published by the registered company Springer Nature Switzerland AG
The registered company address is: Gewerbestrasse 11, 6330 Cham, Switzerland

Preface

Open-source hardware and software technology is transforming numerous fields, and biomedical engineering is no exception. In recent years, a growing number of initiatives have paved the way for open biomedical engineering, democratizing access to tools and knowledge that were once confined to specialized institutions. This book aims to provide a comprehensive end-to-end exploration of this evolving field, covering everything from hardware and software design to data acquisition, including also processing tools, data storage, and applications.

Open biomedical engineering is a rapidly advancing field, and with this book, the goal is to offer an integrated and practical approach to its key concepts. While open-source biomedical solutions are increasingly adopted by students and professionals across multiple domains, essential knowledge is often fragmented. This can be overwhelming and confusing, especially nowadays where there is a plethora of information sources.

While many existing books provide an excellent theoretical foundation for specific topics, often they lack the hands-on, practical guidance required to apply these concepts effectively. The inspiration for this book stems from firsthand experience gained through teaching and lecturing; each year, approximately hundreds of students participate in university-level courses and workshops focused on the addressed topics, underlining the demand for structured, practical learning materials in open biomedical engineering.

This book covers the essential topics in an all-round structured manner. Technical content is complemented with real-world case studies and solutions to common challenges encountered by both new entrants and professionals in biomedical engineering, psychophysiology, human-computer interaction, and related fields.

The content is deeply rooted in state-of-the-art advancements, utilizing modern hardware platforms and widely adopted programming languages such as Python for signal processing and analysis. Each chapter is enriched with graphics, source code, or other additional resources to ensure a hands-on and engaging learning experience for the reader.

By consolidating foundational knowledge with practical insights, this book caters to the needs of students, researchers, and industry professionals looking to explore and contribute to the exciting world of biomedical engineering benefiting from low-cost hardware and open software tools.

Lisbon, Portugal Hugo Plácido da Silva

Acknowledgments This work was supported in part by Fundação para a Ciência e Tecnologia (FCT) under grants 2020.06675.BD and 2022.12369.BD (DOI: https://doi.org/10.54499/2022.12369.BD), Ministério da Ciência, Tecnologia e Ensino Superior (MCTES) through National Funds and Co-Funded by European Union (EU) Funds under Projects UID/50008/2025 - Instituto de Telecomunicações (DOI: https://doi.org/10.54499/UID/50008/2025) and LA/P/0109/2020 (DOI: https://doi.org/10.54499/LA/P/0109/2020). Furthermore, this work was co-supported by the European Regional Development Fund (FEDER), through the Lisbon Regional Programme (LISBOA 2030) of the Portugal 2030 framework under Project LISBOA2030-FEDER-01318700 (ComSense).

Contents

Structured Methodologies for Iterative Concept Design and Biomedical Device Development

Ricardo Baptista

1 Introduction

By definition of the World Health Organization (WHO), Council Directive 93/42/EEC, or ISO 13845, a medical device is:

> (a) [...] any instrument, apparatus, appliance, software, material or other article, whether used alone or in combination, including the software intended by its manufacturer to be used specifically for diagnostic and/or therapeutic purposes and necessary for its proper application, intended by the manufacturer to be used for human beings for the purpose of: - diagnosis, prevention, monitoring, treatment, or alleviation of disease, - diagnosis, monitoring, treatment, alleviation of, or compensation for an injury or handicap, - investigation, replacemen,t or modification of the anatomy or of a physiological process, - control of conception and which does not achieve its principal intended action in or on the human body by pharmacological, immunological or metabolic means, but which may be assisted in its function by such means;

while a medical device accessory is:

> (b) [...] an article which whilst not being a device is intended specifically by its manufacturer to be used together with a device to enable it to be used in accordance with the use of the device intended by the manufacturer of the device.

Therefore, the major goal of a medical device is to improve populational well-being by increasing diagnostic performance, preventing an illness or its symptoms, or compensating for a lost bodily function due to injury or handicap. A medical device may have an active role in the improvement of well-being, for instance, by using an external fracture fixture to heal a large bone fracture. This type of medical

R. Baptista (✉)
UnIRE, Instituto Superior de Engenharia de Lisboa, Instituto Politécnico de Lisboa, Lisbon, Portugal

LAETA, IDMEC, Instituto Superior Técnico, Universidade de Lisboa, Lisbon, Portugal
e-mail: ricardo.baptista@isel.pt

H. P. da Silva et al. (eds.), *Open Source Biomedical Engineering*,
https://doi.org/10.1007/978-3-032-03655-1_1

device enables a faster healing process (some external bone fixing apparatus can induce micro deformations, increasing bone formation speed) while allowing the patient to stay active during the recuperation phase. Or it may have a latent effect, like using a glucose monitor to control a patient's blood sugar levels over a long period of time. This medical device can be combined with an insulin pump, enabling the patient's glucose levels to be monitored while insulin can be administered automatically. Therefore, the long-term side effects of diseases like diabetes can be minimized [1].

The success of a medical device as a product should not be viewed solely from the patient's perspective. While monitoring or controlling illness signals and symptoms can improve a person's well-being, diagnosis and treatment are ultimately carried out by healthcare professionals. Medical devices also improve the clinical procedures of physicians and nurses. These products help doctors reach the right diagnosis in a shorter amount of time, help them to prepare for a surgical procedure (by using simulation software and patient image data), and perform that procedure by using advanced and customizable medical devices. After recovery, one must not forget that the patient will return home, where medical devices will be used by the patient or by nonprofessional caregivers. In the home environment, medical devices are essential and must have high tolerance for incorrect use and be error-proof. Nonprofessional caregivers will most of the time have little experience while using these devices, and there is a high probability of making mistakes while using them or reading the results on a medical device display. It is very important to develop error-proof medical devices for home use [2].

This is one of the main reasons structured methodologies play such a crucial role in medical device development and concept generation. Products, like medical devices, should be developed with the end user or customer needs in mind. In terms of medical devices, the end user or possible future customer can be a patient or a health professional. But if a product is not developed with their needs in mind, then it will not satisfy end user or customer necessities, rendering the product unsuccessful. Structured methodologies start with the identification of a market opportunity, where a patient's or a health professional's needs are still not fully satisfied or can be fulfilled in a more efficient way. Then a new product concept will be designed, prototyped, and tested. If this new and more innovative product performs better than the competition, it can be produced and sold, improving the population's well-being. As the process started by identifying the customer needs, the whole process is centered around these necessities, improving the product performance. Structured methodologies also provide advantages like (a) quality assurance, by establishing project milestones; (b) project coordination, by specifying the responsible for each task; (c) project planning, by defining the task order; (d) project management, by providing documentation for all project phases; and (e) improvement, enabling the team to identify and learn from previous mistakes [3].

Structured methodologies are also a source of improvement, trust, and creativity within a biomedical company, to develop new and more innovative products [4]. Many authors consider innovation the key to better product development. In the case of medical devices, Durfee et al. consider innovation the way to add value to society. To achieve this goal, health professionals, like physicians, should start the development process by making a sketch for an idea that will solve a medical problem. If this idea is transformed into an innovative product, then it will be better than its competitors [5]. Some ideas may require a simple product function to be successful, while others may be more complex. In this case, functional decomposition plays a major role in concept and product development [6–8]. In this process, each product function is separated into a simpler subproblem that can be easily addressed, facilitating concept generation. By assigning a product physical component to each function, final product architecture can be defined. In the case of medical devices, product function is, in many cases, correlated to clinical procedures. Hagedorn et al. developed a model for functional decomposition of medical devices and ideation of new concepts. This model integrates function with clinical practice and engineering to provide innovative and safer solutions for medical devices [9].

Although there may be different structured methodologies, medical device concept and product development are multidisciplinary tasks. Therefore, several authors consider it essential that a medical device is developed with a multidisciplinary team, where not only engineers and designers, but also health professionals must be present. This is then a collaborative process, where communication is essential between different disciplines. Vincent et al. consider the different languages spoken by different professionals the first development barrier to be overcome [10]. Multidisciplinary teams are also very important because they allow designers to see how a product is used in clinical practice [11, 12]. Martin et al. consider that clinical practice should be the push for new technology development, and it is up to biomedical companies and the healthcare sector to allow designers to observe real-world use of existing medical devices [11]. Today Human Factor Engineering (HFE) defines the interactions in a user-device interface [1]. When not properly developed, this interface may cause errors, especially in the home environment, where medical devices are used by nonprofessional users [2]. Telemedicine is also on the rise and can solve some of these problems, but when used by older adults, HFE plays a very important role in product development [13]. Understanding the basic principles of HFE is one of the most complex components of medical device development. But it will enable developers to achieve safer and more efficient products, even when misused by users. Users will use a medical device to do what they want, in the way they expect it to [14]. Therefore, Branaghan et al. define usability as a synonym of HFE and consider it to be defined along five dimensions: (a) learnability, (b) efficiency, (c) memorability, (d) error resistance, and (e) satisfaction. To include these dimensions on the development process, it requires early user observation and a multidisciplinary team that is able to take advantage of this evidence-based design [15]. Gurses et al. consider the development of HFE specialists and their incorporation in the medical device development process to be fundamental. In order to do this, HFE courses should be included in physician, nurse, and engineering education, promoting HFE workshops or creating HFE master's and doctoral training programs [16].

Examples of multidisciplinary teams working with structured methodologies on medical device development can be easily found. Sack et al. present a methodology for including pediatric surgeons on medical device development teams in order to create innovative products, designed and customized for pediatric surgical interventions. These health professionals played an effective or even leading role on a team of engineering students and professors, enabling the development of medical devices customized for the different patient anatomy. Therefore, surgeons no longer need to adapt procedures to the use of medical devices designed for adult patients [17]. Ranger at al., on the other hand, used structured methodologies (design thinking) to develop a sustainable and cheaper prosthetic device for the developing world, combining on the same team engineering students, doctors, patients, and international stakeholders. Several different concepts were achieved, and some students actually continued this research beyond the scope of the training exercise [18]. Therefore, one final conclusion is that structured methodologies also play an important role in engineering and health professionals' education, leading to a future where medical devices are developed from the ground up, with the patient's needs or clinical procedures practice as the major focus points.

2 Concept Development

The product development methodology presented in this chapter is similar to the one developed by Ulrich and Eppinger [3], Yock et al. [19], or Ogrodnik [20]. It represents a structured methodology from opportunity seeking to final product commercialization. As a structured method it is based on different steps: (1) opportunity seeking, (2) planning, (3) concept developing, (4) system design, (5) detailed design, (6) testing and certification, (7) production, and (8) commercialization (Fig. 1). This is not a rigid structure, as biomedical companies, health institutions, and education and research institutions should adapt this structure according to their needs. Opportunity seeking is essential for the business success of biomedical companies. While health and research institutions typically begin the product development cycle directly from the concept development phase, as they are trying to respond to a specific market need, detailed design, production, and commercialization are generally carried out by biomedical companies, as educational institutions usually lack manufacturing capabilities. Therefore, planning, concept development, and system design are fundamental for all types of product developers.

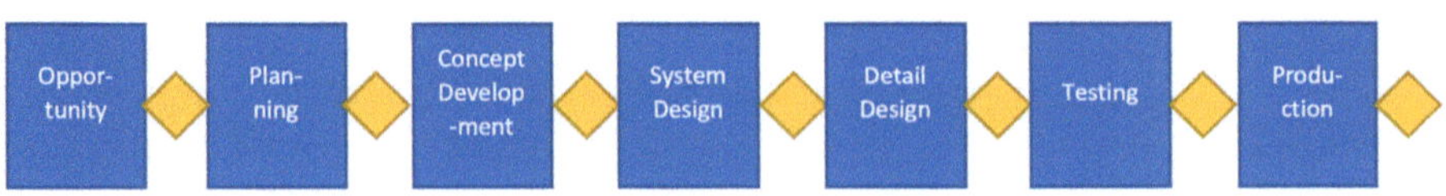

Fig. 1 Structured methodology for product development, from opportunity seeking to final product commercialization [3]

Opportunity-seeking activities enable developers to find new ideas for biomedical products. The driving force behind this activity can be a patient's need or a health professional's need, identified by the biomedical company. But it can also be a newly developed technology that can be applied to solve biomedical problems more efficiently. Once identified, this opportunity must be assessed, as there are always risks involved when developing a new idea. These risks can be divided into different categories, like economic or technological risks. Biomedical companies should ask themselves questions like: "How well do we understand the technology likely to be used?" "How well do we understand the patient health problem?" "How well do we understand the market for this new idea?" Answering these questions is fundamental to understanding the impact and the possible risks for the company when developing a new idea. Will it lead to the development of a new product based on existing technology, with minor improvements and low cost? In this case, the new idea shows low financial risk for the company. But if a new idea leads into unknown territory, for both technical and market dimensions, then it will result in a more innovative product with a higher economic risk for the company.

Many opportunities should be explored in order to generate a viable idea for product development. At the end of this stage, a new idea should be represented by a simple description and a sketch for the future product. Some companies will also write their mission statement. This document represents the direction where the new product will take the company, without specifying the final destiny or how the product will solve the problem. The mission statement is a working base for future activities and is comprised of the idea description, the benefits for all stakeholders, the product goals, market(s), and general characteristics.

The planning stage depends on the type, size, and structure of the product developer, and it is out of the scope of this chapter. At this stage, companies define the human and technical resources to be allocated. One should be aware that product development is a multidisciplinary activity. A development team should have a leader and specialists from all the involved areas of expertise: biomedical engineers, human factors engineers, health professionals, designers, industrial designers, marketing professionals, purchasing specialists, end users, etc. When developing a biomedical product, one of the major goals is to reduce the likelihood and possible consequences of error. Bitterman et al. consider the collaborative work between product developers and end users vital for product development success [2]. The end user of a biomedical device can be a health professional, normally using the device in a hospital environment, where there are multiple gateways checks and multiple users checking the same information. This is a way to reduce the possibility of error occurring. But the end user of a biomedical device can also be the patient, a caregiver, or a health professional in a home context, where the device is used in an individual fashion. In this case, product design is fundamental to reduce the likelihood of error, because of the reduced number of gateways and opportunities to check if the information given by the biomedical device is correct. Including the end user in several product development stages plays a major role in the success of this methodology. In many cases, the end user suffers from cognitive and motor disabilities, as a consequence of their age or chronic illness. This limits

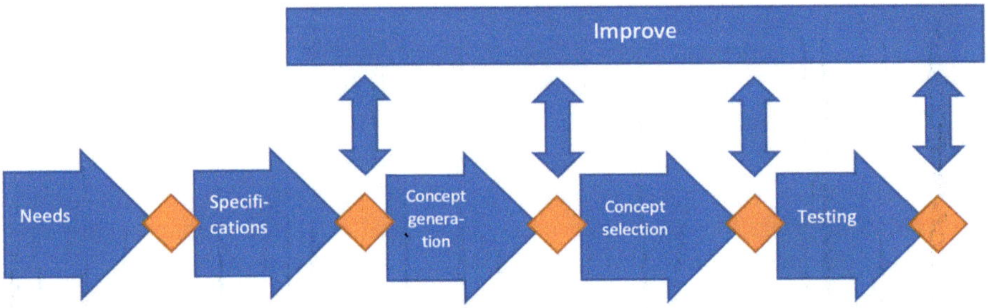

Fig. 2 Concept development stage, from user needs to final product specifications

their performance and ability to correctly use a biomedical device at home [11]. Today, it is considered fundamental to have human factors engineers on the product development team. They are the key to a successful product development and the interface between developers and users [10].

Concept development can be considered the most important stage in product development. A concept represents the way a new idea will solve a biomedical problem. It only represents a model, not the full detailed design that will be later on produced and sold. But without a good concept, there cannot be a good product. Concept development is the focus of this chapter and can be divided into six different steps: (a) user needs, (b) target specifications, (c) concept generation, (d) concept selection, (e) testing, and (f) final specifications (Fig. 2). All these steps will be fully detailed in this chapter.

On the system design stage, the final product architecture is defined. The architecture defines the basic mode of operation and how functionality and physical components are related. This structure has already been partially defined on the concept generation step. This makes products modular and not integral (Fig. 3a), and the type of modularity chosen by the biomedical companies plays a major role in their economic success. Products can be defined using a slot architecture (Fig. 3b), where different components have different system interfaces. This is the most common product architecture, as companies normally work with different suppliers and do not control the development of all the product components. The product will be cheaper to produce, but the final performance is affected and limited by the different interfaces. With a bus architecture (Fig. 3c) all components share the same interface, but they all connect to a central bus. This architecture requires the development of a central component, controlling all product functions, increasing product performance and also the final cost. Finally, a sectional architecture (Fig. 3d) is also possible, where all components share the same interface and can be connected to each other. This is a serial architecture, reducing the cost, but limiting the communication between product components. Product architecture influences product performance, but also company economy [6].

Detailed design includes all the activities necessary to achieve the product's final technical drawings and material selection. This stage will influence the manufac-

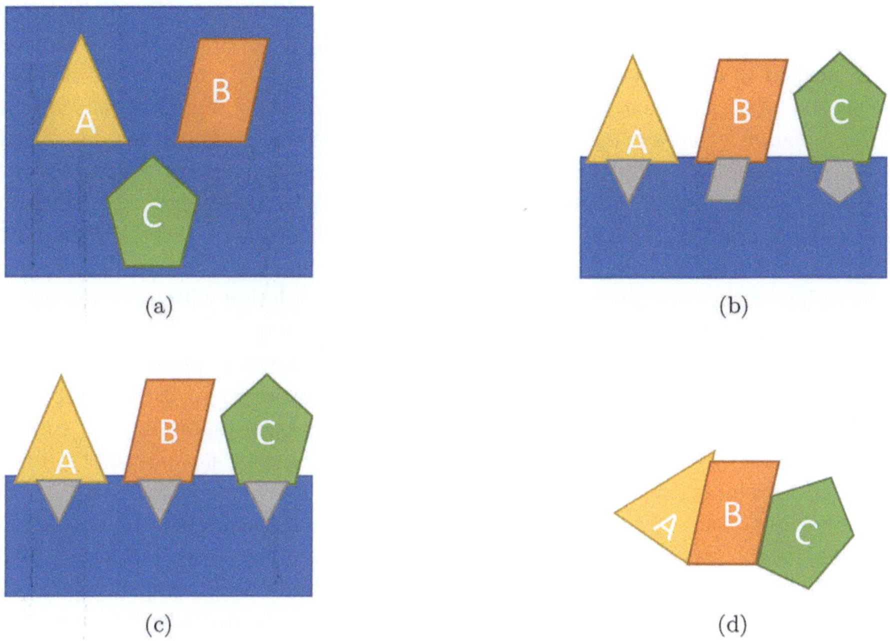

Fig. 3 Product architecture types: (**a**) integral, (**b**) slot, (**c**) bus, and (**d**) sectional [3]

turing technologies and can be subjected to an iterative approach until the final details are achieved. Small decisions like the interface between different product components can have a major influence on the final product. If a coupled interface is chosen, then this interface depends on the component geometry and materials. Altering one component can lead to costly interface redesigns and make product updating a very complicated task. This will lead to underperforming products. Using an uncoupled interface means that components can be changed or updated with ease, but interface design can be harder [6].

Testing and certification are fundamental tasks before final production ramp-up. Testing can be done initially with prototypes, assessing product functionality. Certification is mandatory before introducing a biomedical device to the market, and procedures depend on the geographic region in question. Certification will be discussed later in this chapter. Finally, production and commercialization lead the product to the market, but these two topics are out of the scope of this chapter. The previous methodology is clearly adapted for the development of market-pull products. A biomedical company identifies an opportunity, most likely based on an end user need, and develops a new product based on existing technologies.

Biomedical companies can also develop other types of products. Technology-push products begin their life with an opportunity that arises from the development of a new technology or material. These new technologies or materials are sometimes developed outside of the company, by research centers or higher education institutions. They are then transferred to an external company, or a startup company is created around these ideas, when an application is discovered. These products are then pushed to the market by technology development. Many biomedical devices were developed as technology-push products; for example, a digital stethoscope is a combination of an analog stethoscope and telemedicine technologies developed with other goals. By pairing an old concept with a new technology, a new product was developed, and its advantages were applied to improve medical education and clinical practice. The product development methodology is very similar. The mission statement can already include the chosen technology, and on the concept generation stage, the main goal is to compare the new solution with its competitors, as the product will only succeed if it offers a clear advantage.

Technology-push products can also be used to develop platform products. Biomedical companies can invest large amounts of resources in the development of one specific technology. Once this technology has been market-tested, similar products with different applications or specifications can be developed. Platform products development is very easy and inexpensive.

Process-intensive products, on the other hand, are constrained by a specific manufacturing process. Their design cannot be separated from the production technologies, even on the concept generation stage. Biomedical companies start by choosing or creating a new manufacturing process before the mission statement, and all the development methodology is based on this constraint. The resulting product is therefore produced on a large scale in order to be economically viable. Examples of process-intensive biomedical products are dental implants. Their development and design are limited by the manufacturing process and strict certification rules, in order for it to be cleared for clinical application.

Customized products are the opposite of process-intensive products. They are designed, developed, and manufactured at a small scale in order to satisfy a customer-specific need. Companies can use their own technology with the necessary modifications to design and produce customized products. Examples of biomedical customized products are leg or arm prosthesis and ankle-foot orthosis designs for a single customer. Normally, the development process starts with image acquisition, by laser scanning, followed by a concept generation stage where a tridimensional model is created, and the small-scale production can be performed using 3D printing technologies. The final product price can be higher than process-intensive products, be the performance of a customized product its biggest advantage. Finally, complex systems products in biomedical applications must be developed using different strategies. These products must be divided into many subsystems, and therefore, product architecture plays a major role in product development. System functionalization means that many product functions will be developed at the same time. The development team will be very large, and coordination is fundamental. Examples of complex systems in biomedical products are health equipment like a computerized tomography scanner or a magnetic resonance scanner. These include many subsystems with specific functions. For each system, a different technology must be developed in parallel.

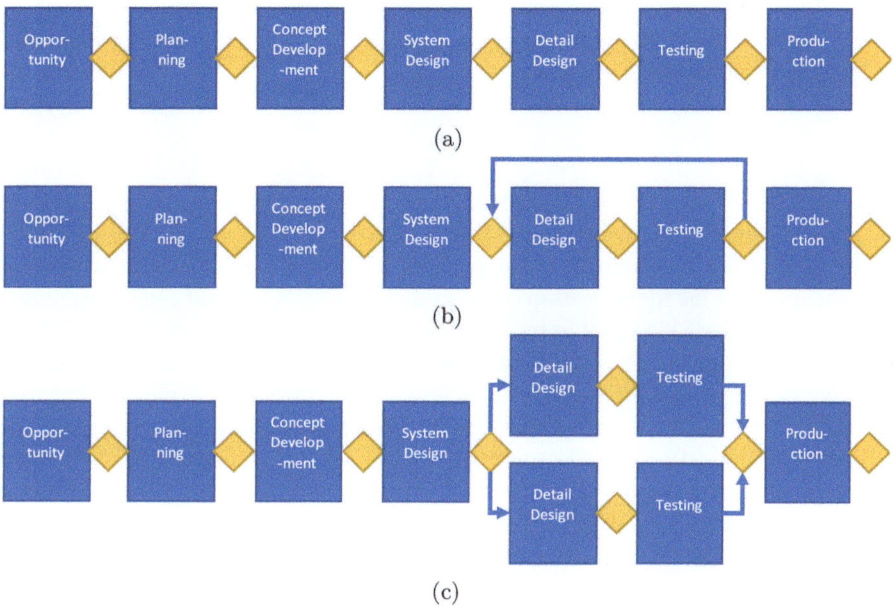

Fig. 4 Different product development flows, (**a**) linear, (**b**) spiral, and (**c**) parallel [3]

The product development flow can be altered accordingly to the product type. Market-pull, technology-push, platform, process-intensive, or customized products are normally developed in a linear fashion. Each step is followed by the consecutive step, while a gate between them allows for quality assurance (Fig. 4a). A spiral flow (Fig. 4b) can be set up in order to accelerate product development. Many iterations can be made in both the concept generation and selection stage and the detailed design and testing stages. Technologies like 3D printing are very important for this type of product development flow, allowing for quick concept or design prototyping and testing. Complex system products must be developed using a parallel flow, where the product is divided into many subsystems at the system level design stage (Fig. 4c), and these systems are then developed in parallel to accelerate product development.

Using a structured methodology for product development offers clear advantages when compared to nonstructured methodologies. Establishing milestones between each development stage, defining the conditions that allow the project to move on, is a quality assurance method. Structured methodologies also provide a coordination tool for the project leader, as each person on the team can have a clear task assigned to them. These tasks are also representative of planning tools, making project management easier. Finally, structured methodologies provide documentation for the project; therefore, team members can use this information to improve their performance on future projects.

As mentioned, the concept development stage is of major importance for product development. Special attention will now be dedicated to the several steps comprising this activity.

2.1 Identifying End-User Needs

When developing market-pull or customized products, the concept development stage begins with the identification of the end user's needs. A need can be expressed using the end user's words, it is independent of the product a biomedical company wants to develop, and it should be assessed independently from the technical solution used to solve the end-user problem. In fact, when determining end-user needs, developers do not know what solution will be used or even if the necessary technology to satisfy the end-user need already exists. A need is than a future customer desire, and it can be expressed as a product function that solves a specific problem. As in so many areas, the biomedical field is full of potential problems that need to be solved and where innovation can lead to the development of new products that will better satisfy customers' needs.

The starting point for identifying end-user needs is opportunity seeking. Developers can observe real people in real-world situations. Using medical devices in everyday clinical practice, in order to understand if there is room for improvement. Economic challenges are very important in medical practice; understanding the procedures is the first step to understanding how to make clinical practice more sustainable and safer. This is another source of opportunities that involves multidisciplinary teamwork and one of the reasons why health professionals and customers or end users should be part of the development team from the start. The end user can be a surgeon who needs to perform a complex procedure with only one hand; therefore, a new medical device must be developed, or a patient in his home, who needs a biomedical device to help him eat autonomously. Identifying end-user needs should be a structured process in itself. There are four main steps in this activity.

0. (Opportunity seeking).
1. Gather data from the end user.
2. Interpret the data and construct the end-user needs.
3. Organize the end-user needs.
4. Sort the end-user needs by importance.

Structuring this activity is fundamental for product development success. It will ensure the focus of the product on the customer's needs. If a company develops a new product based on the wrong necessities, this product will not satisfy the end-user needs; therefore, this product will not be adopted and will ultimately fail. This methodology can also identify hidden needs, allowing the development of better-performing products and increasing the overall product quality. It will also justify all the product specifications and generate documentation, as we will see in the next item.

But probably the most important feature of this structured procedure is that it enables communication between developers and end users. Developers will be able to experience the end-user needs firsthand by working directly with the product consumers. As mentioned, human factors engineering must be an integral part of biomedical product development. Therefore, developers must clearly understand how end users will use the product, and that is only possible to achieve with developer-end-user interaction. Branaghan et al. consider usability, multidisciplinary interaction, and the use of structured methodologies fundamental for safer biomedical device development [15].

Gathering raw data from the end user can be performed in several different ways. A large company should use different approaches to optimize the information gathering process. Personal interviews are the basis of the gathering process. One or two developers should sit down with a potential user or customer. The interview should be carried out in the user's working or living environment, so developers can also explore this source of information. An interview should not be longer than two hours, and a list of specific questions should be prepared in advance. The interview normally starts with open-ended questions, and the answers should not be influenced by the interviewer. This will maximize the different types of needs obtained by the interviewer. Then the interviewer can start to use more closed questions, directing the interview and obtaining more detailed needs, which will later be translated into product specifications. Online interviews and questionnaires should be avoided, as no interaction between developers and end users will happen. A few examples of questions that can be part of any type of interview are:

(a) When will you use this product?
(b) Why will you use this product?
(c) How will you use this product? Describe all necessary steps, please.
(d) What do you like about product a?
(e) What do you dislike about product b?
(f) How would you improve product c?
(g) What features do you consider when acquiring this type of product?

Focus groups are more complicated to set up. In a focus group, between 8 and 12 end users or possible customers are in the same room, discussing their needs under the moderation of one developer. The focus group can be recorded, or the rest of the development team can watch the debate live in a different room. From the interaction between participants, many hidden needs can be determined, but some fine details can be lost in this type of gathering process. Finally, the gathering process can be performed by watching competing products being used. Or when the need arises from a missing product feature, developers should fully understand the procedure in question to gather information about end-user needs.

Gathering information can be an expensive and time-consuming activity. Selecting the right end users to interview is essential for increasing the process efficiency. Interviewing and focusing on lead and extreme users is one possibility. Lead users are years ahead of the market. They experience the product needs well before other users, are able to explain their needs in a more detailed fashion, and sometimes have already identified their own solutions for the problem. Extreme users have special needs (and sometimes also their own solutions), and their opinion will lead to the development of better-performing products, even under extreme conditions. Working with these types of users will result in the development of more innovative products, and these innovations and new technologies can later be integrated into normal products. Interpreting the collected data is the process where the end users' or customers' needs are truly constructed. Data from the interviews, focus groups, or observations is written with the customer's own words. Different users will also use different languages to describe the same procedure or the same problem. In this step, developers must interpret the data and write the needs using a common language. This task should be performed by different developers, initially individually and secondly as a group. The most commonly used technique is to translate the raw data into small pieces of paper. Each identified necessity is written in a different piece. Then, working in a group, developers can identify identical needs and organize them. Use a large board for this task, and no need will be forgotten. The need should be written using positive phrasing, as this is better for specification translation. Needs should be as detailed as the raw data; no information should be lost in translation. Needs should be written as product attributes or functions, making it easier for the specification translation. Needs should specify what the product must do and not how the product will do it. End users sometimes already have an idea how to solve the problem, but what is more important than the how at this stage.

The third step of the end-user need identification activity is to organize the translated needs. Again, a large board (Fig. 5) is a major helping factor in the step.

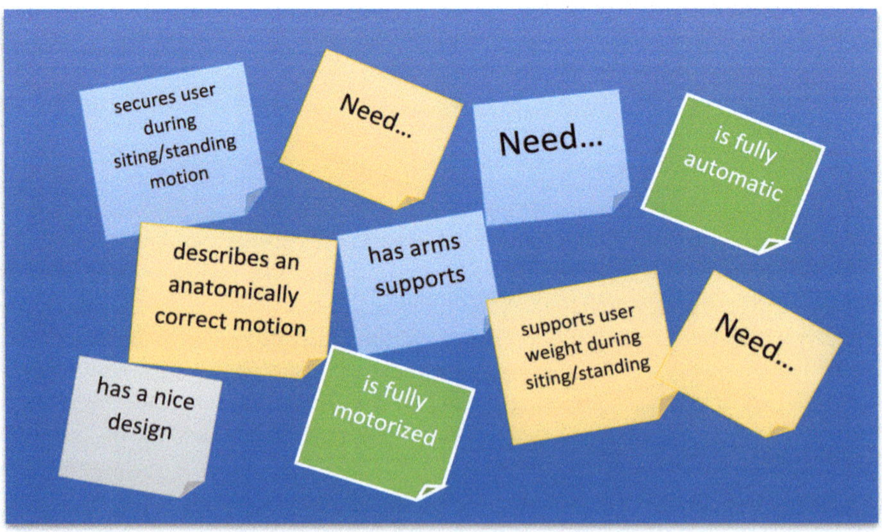

Fig. 5 Example of a visual exploratory method for organizing end-user needs, some of the possible needs found for a lift chair medical device

Table 1 Example of an organized and classified by importance needs list, for a lift chair

Need: the lift chair...	Importance
Secures user during siting/standing motion	4
Has arm supports	3
Describes an anatomically correct motion	4
Supports user weight during sitting/standing motion	5
Is fully motorized	2
Is fully automatic	2 (!)
Has a nice design	2

Developers can organize the pieces of paper visually on the board with ease. They should organize the needs by type and hierarchically. A type of need is connected to a product function. Depending on the complexity of the product, more types of needs can be defined. A hierarchical needs list is a set of primary needs, secondary needs, and, for very complex products, tertiary needs.

Finally, on the last step, the organized needs are sorted by importance. It is very hard to sort all needs numerically, as 50 to 300 needs can be determined for a single product. Therefore, needs can be sorted using a 1 to 5 scale, where 1 is an undesirable feature, 2 is a non-important feature, 3 is a basic feature, 4 is an important feature, and 5 is a critical feature. It is also important to identify special needs that will be translated into advanced product features, differentiating the product from the competition. These needs can be marked with a "*" or "!", to be quickly identifiable. In the end, a fully organized end-user needs list (Table 1) will be available for the team to pursue the product development plan.

2.2 *Defining Product Specifications*

When considering the end-user or customer needs, one must consider two important disadvantages. Developers must take these into account when developing a product; therefore, biomedical companies generally prefer to work with specifications or requirements. Usually, when developer finalize their needs list, they are:

- Written in a language close to the one used by the end-user or customer
- Difficult to quantify

Therefore, a product specification or requirement should be measurable and written in a more technical language. A specification tells the developer team what the product must do, in a precise, quantifiable way. A need will be hard to verify because it deals with subjective questions. A specification can be measured; therefore, once the product prototype is built and tested, developers can easily check if a specification is met. For each specification, a metric should be defined. A metric is the numeric value that quantifies the specification. Using a single value for one metric will lead to difficult situations, as it will be hard for a product to meet that specification. Developers usually use intervals, "less than..." or "higher than...," to define the specification. After testing, if the metric falls within the defined interval, then the product meets the specification.

In a perfect world, the initial specifications would be set definitively, through all the product development cycle. Unfortunately, in the beginning it is very hard to know exactly the real values for specification metrics. Therefore, before concept generation, selection, and initial prototyping, these are just target specifications. They will be very important for testing and product benchmarking against competitor products. But as the project unfolds, many trade-offs and constraints will change the specifications' metrics. For example, the desired material to manufacture a stethoscope bell might be available, and the production technology exists within the company, but the final production cost is not viable for the project's success. The development team decides to exchange the material with a cheaper one with a higher density and different acoustic transmission properties. As a result, the final weight and acoustic performance metrics of the product will have to be altered. Trade-offs and constraints lead to the final specifications at the end of the product concept stage.

The goal is then to translate each end-user need into one specification and associate this specification with a single metric. This would put the end user or customer at the center of the process. The product would be developed focusing on the customer and would fully satisfy the customer's needs. In reality, this may not be possible. The developer team might not find on metric for each need, or some needs might require more than one metric to be fully assessed. In order to structure this process, a needs-metrics matrix should be prepared (Table 2).

In each Table 2 row, the developer team will find the end-user needs, while in each column, the team will find a product metric. The association between a need and a metric is marked on the corresponding cell. Ideally, this would be a diagonal matrix. Considering the lift chair example, Table 3 shows the defined product metrics. Analyzing both Tables 2 and 3 it is possible to see that while securing the user has been defined by the minimum chair dimensions or the existence/nonexistence of arm supports, the need for an anatomical movement was defined by the minimum number of mechanism bars. Weight support and motorization were translated into one single metric, having a motor with at least 2000-W power. Finally, being fully automatic means the lift chair should have an integrated controller and remote control.

The needs metrics is the basis of the House of Quality (Fig. 6), a graphical approach technique for customer-centered product development. The House of Quality is part of the Quality Function Deployment (QFD), a process that enables developers to choose the appropriate trade-offs between the customer desires and requirements and what the biomedical company can technically and economically offer [21]. On the House of Quality construct, the end-user needs are represented as the "What's" (on the left side of the house), while the specifications are called the

Table 2 Example of a needs-metrics matrix, for a lift chair

Need: the lift chair...	Chair	Arms	Mechanism				
Metric	dimensions	supports	number of bars	Motor power	Controller	Remote	Designed by...
Secures user during sitting/ standing motion	*						
Has arm supports		*					
Describes an anatomically correct motion			*				
Supports user weight during sitting/standing motion				*			
Is fully motorized				*			
Is fully automatic					*	*	
Has a nice design							*

Table 3 Example of values to be attributed to different metrics, for a lift chair product development

Metrics	Value
Chair dimensions	500 × 500 [mmxmm]
Arm supports	Yes/no
Mechanism number of bars	At least 4
Motor power	2000 [W]
Controller	Yes/no
Remote	Yes/no
Designed by...	Yes/no

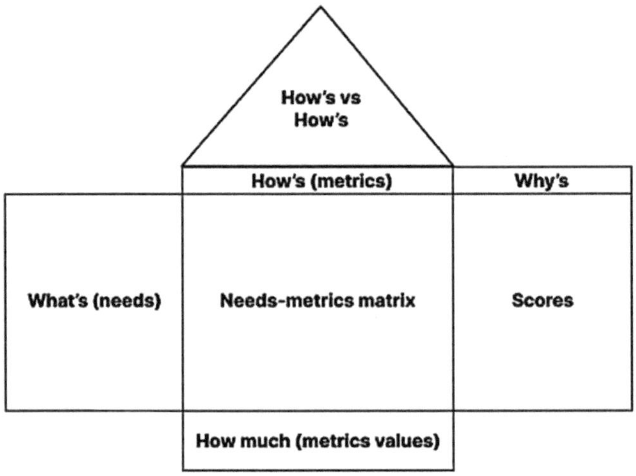

Fig. 6 Example of a QFD matrix [21]

"How's" on the top of the house. Finally, the metrics refer to "How much" at the bottom of the house.

As mentioned, metrics should be dependent variables, like the weight of the stethoscope bell. The weight depends on the material density; choosing the material as a metric will constrain the project. Therefore, developers should only choose dependent variables as metrics and set the appropriate interval for that metric. Choosing independent variables as metrics leads to fewer options on the concept generation step, and the quality of the final product will decrease. The interdependency of specifications can also be represented in the attic of the House of Quality. These relationships can be considered positive or negative, weak or strong, and different developers use different symbols to represent these relations on the attic of the house [21]. A positive relationship means that by improving on one metric, other metrics will also be improved, while a negative correlation means that altering one metric, another will be harder to reach.

Some end-user needs may be too subjective to be translated by a single number. These are classified by Hatz et al. as power features, on the Physicians' Motivation to Adopt scale, and can be evaluated by a panel of customers. They are not quantifiable using physical units, but rather classified using different scales. Power features are connected to motives like recognition, self-determination, or superiority, describing customers' desire to improve their social and professional position [22]. Finally, metrics are also fundamental for product benchmarking. Each end-user need can now be compared against competitor products, using defined specifications. This can also be used to check if the chosen metrics are viable and can realistically be achieved. In the House of Quality, competitive benchmarking is a fundamental task. On the right side of the house, the "Whys" can be added. Different competitive products are selected, and a comparison can be made using the end-user needs as a reference. Products can be rated from poor to very good, regarding the perceived satisfaction of each need, and therefore, benchmarking is focused on the customers' opinion [21].

2.3 Concept Generation

A concept represents the development team's idea of how to solve the end-user needs. The concept describes the technology, working principle, and the general form and design of the future product. Concepts are generally represented using a simple sketch, but nowadays it is also possible to use computer-assisted drawing tools to create virtual 3D models to represent different concepts. Virtual 3D models can be prepared using free software tools and can even represent the working principle of the concept, illustrating the physical relationships between different parts. Virtual 3D models can also include and represent the different materials to be used, making them a very useful tool. Finally, concepts can also be represented by a 3D printed model, making concept selection an easier process. Biomedical companies should also take into account that concept generation is a relatively inexpensive activity when sketching concepts. Creating virtual or physical 3D models will take more time and, of course, be more expensive.

To generate concepts, developers start with the end-user needs, product specifications, and metrics and individually create solutions that satisfy these conditions. The goal of this individual work is to increase the number of ideas and concepts generated. Failure to consider more than three or four ideas from each developer, failure to consider ideas from different developers, failure to consider how other companies already solve the problems, or even failure to consider ideas that initially seem ineffective, will most certainly lead to poor concept generation and poor product performance. Therefore, once again, the development team should use a structured methodology to complete this task. Generally, developers should:

1. Decompose the problem
2. Search solutions internally
3. Search solutions externally
4. Explore and create new solutions

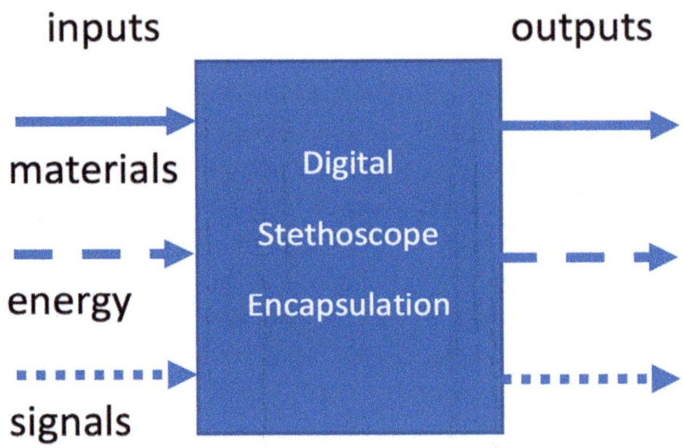

Fig. 7 Initial black box product decomposition, example for a digital stethoscope encapsulation

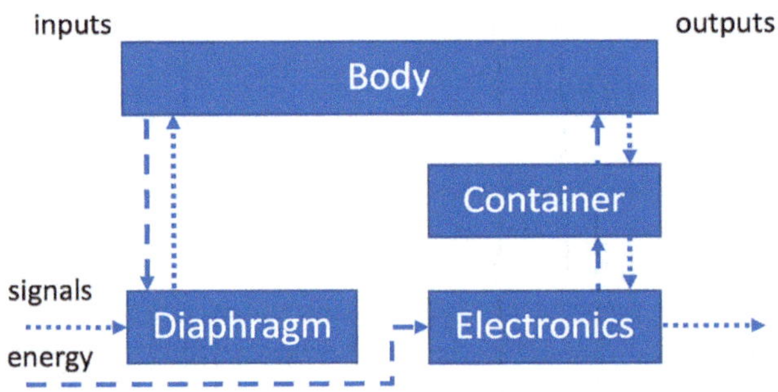

Fig. 8 Product functional decomposition, example for a digital stethoscope encapsulation

Decomposing the problem is considered an essential task for concept generation [6]. The goal is to clarify the problem by decomposing it into simpler subproblems. Companies normally use functional decomposition, as problems are decomposed using inputs and outputs. Initially, a product is considered a single black box, with materials, energy, and signals as inputs and outputs (Fig. 7). The next step is to consider three subproblems: clarifying how the product will process energy, materials, and signals and how these functions are related to each other. On a third step, more subproblems are created as necessary, until a full function diagram has been created (Fig. 8). This diagram represents the relation between functions and subsystems, but it does not include the technology to be used yet. Function diagrams are not unique, so development teams must agree upon the final version, as this activity can take a long time to complete.

Another possibility is to decompose the problem accordingly to a sequence of user actions. Hagedorn et al. consider this approach to be fundamental for the development of biomedical products. On their ideation model, function is combined with clinical procedures in order to decompose the problem into different actions. Using a structured framework, it is possible to combine medical knowledge with engineering design to create a product concept based on the underpinnings and reasoning of clinical practice [9].

Once the team has decomposed the problem into simpler subproblems, it is easier to start discovering solutions for each individual problem. At this stage, it is very important to start with the most important function of the product. Team members should decide which function is the most critical, deferring the solution of less important subproblems for a later time. This ensures full attention and focus on the major product function.

The second step of the concept generation activity is to search for solutions externally. This means the team will find solutions that already exist in the market. Using these solutions is quicker and probably cheaper than developing new technologies to solve the problem. It is also a very important step, as it will allow for solution benchmarking once the team has developed their own solutions. The external search can take many forms. One of the most important methods is to interview lead users. As mentioned, these future customers are years ahead of the market and sometimes have already found a solution for their problem; now they might be looking for a partner to implement that solution. Health professionals, like medical doctors, are lead users in their fields of action; interviewing them is a great source of possible solutions for biomedical product concepts. Sack et al. have identified the development of pediatric medical devices as an interesting area where product development structured methodologies should be applied. Pediatric surgeons have struggled with medical devices' limitations, as current practice standards force them to use adult-design medical devices in pediatric procedures. Developing new and children-adapted medical devices has been their necessity for years. Sack et al. developed a methodology to include pediatric surgeons as a main source of external ideas for the development team [17]. Interviewing consultants is also very useful. Searching within universities and specialized consulting companies is essential to find this source of external consulting. Finally, searching on patents' databases and scientific papers, journals, and books is also a very important source of ideas. Scientific papers usually present methodologies and describe solutions for technical problems, while patents include sketches, detailed drawings, and their technical description. Be aware that patents are normally protected for around 20 years. But searching in older patents can be an important source of ideas to generate product concepts (Fig. 9).

On a third step, the development team will search internally for ideas to solve the subproblems. Initially, this is an individual task. Developers must create many solutions for each problem. The recommended process includes a quick sketch or a brief description of these solutions on a simple piece of paper. Then, the team can assemble a large board, the Gallery Method, of different ideas and compare developed solutions together in a brainstorm activity [23]. If the team prefers to use written descriptions instead of simple sketches, then Method 635 is also useful. In Method 635, developed by Rohrbach, six (6) developers write three (3) solutions or describe three possible concepts on a piece of paper, within five (5) minutes. Then these ideas are passed to the next developers, who read these ideas and add three new different ideas to the paper. The circle continues until all developers have contributed [24]. Both methods tend to generate many different ideas.

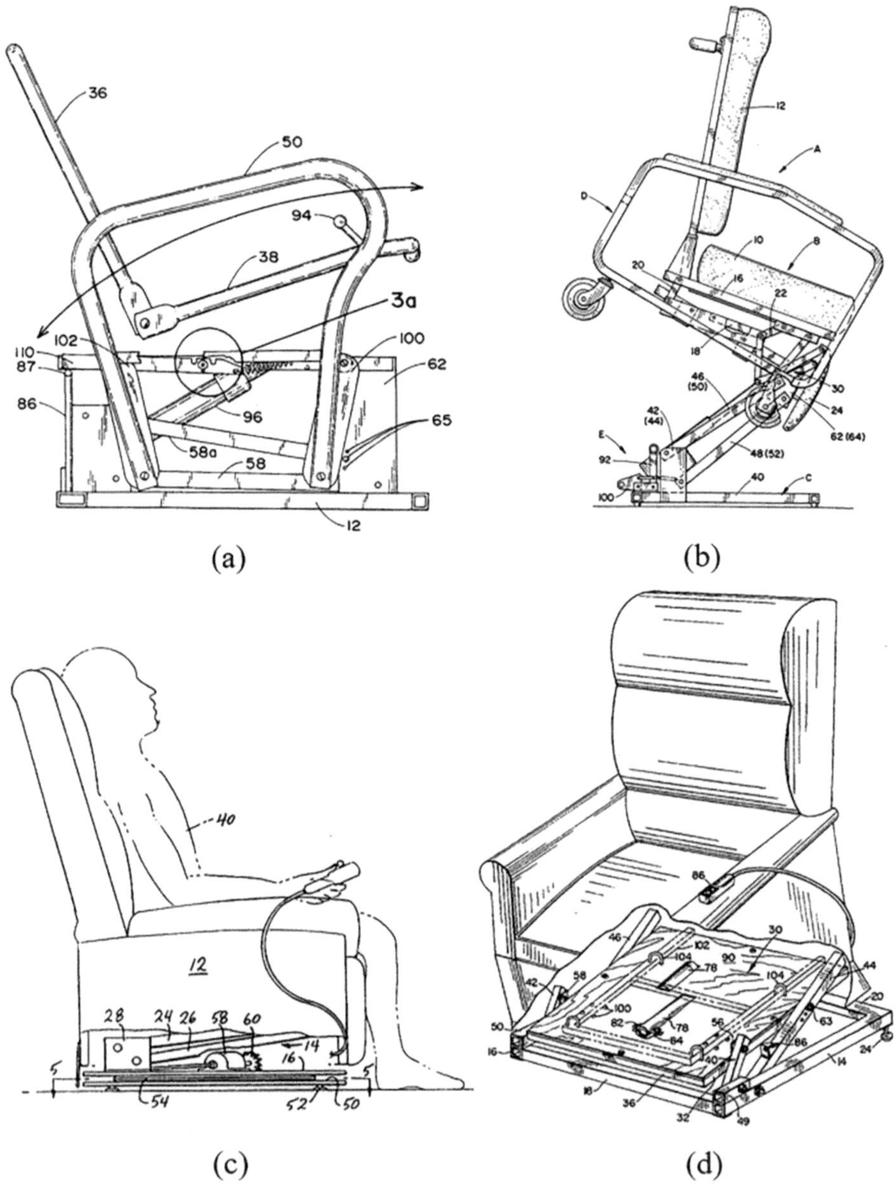

Fig. 9 Examples to patents searched for possible concept solutions for a lift chair, (**a**) patent no., US 6,213,554 B1; (**b**) patent number, 5,203,610; (**c**) patent number, 5,895,093; (**d**) patent number, 5,641,201

This is the most open-ended and creative stage of the development process. The goal is to create as many ideas as possible. Quantity is more important than quality at this stage, and by lowering the quality expectation, more ideas will be developed by the team. Individual criticism of others' solutions should not be part of these meetings. This means that even unfeasible ideas should be considered, as sometimes brainstorming what looks like an impossible solution will lead to a good way to solve the problem. Group work leads to improved ideas, and that is the secondary goal of this stage. Use a visual representation of the developed ideas and the Gallery Method to display a large number of solutions to start the discussion. The team might also use physical representations of the ideas, using common objects. This is useful to use analogies with other products and to use related and unrelated stimuli to create more ideas.

Collaborative sketching (C-Sketch) is also a useful technique for idea generation in the engineering design of medical devices. C-Sketch is a powerful tool for progressive idea generation (Fig. 10), where the whole team collaborates in developing and improving an initial idea, in an iterative form. Provocative stimuli, with the use of analogies and misinterpretations, lead to many design modifications that serve as an origin for new ideas and technologies. While designer feedback with silent criticism leads to the generation of quality ideas, by observing how others interpret each other's designs [25]. This methodology consists of giving everyone a sheet of paper and a pen. All team members sketch a different concept individually. Then, the sheet of paper is handed to the next developer, who will continue to sketch over the other developer's initial idea. The concepts in each sheet of paper will continue to circulate until everyone has contributed. In the end, the resulting concepts will be discussed between the team members. In a simple variation of this method, only one large sheet of paper or whiteboard can be used, where only one concept is generated at a given time, with the collaboration of all team members. This will result in a shared team vision, but fewer concepts will be generated. Shah et al. have found that C-Sketch outperforms both the Gallery Method and Method 635 in the dimensions of quality, novelty, and variety of concepts generated [25].

This stage should end once a specified goal is achieved, for example, the total number of ideas or the total number of hours dedicated to this activity. Finally, the last step of the concept generation step is to explore the obtained solutions for individual problems and to assemble one solution for the general problem. Now that different solutions have been achieved for subproblems, different solutions can be selected for the final product. For example, if one of the subproblems is energy storage, and different solutions were found, these can be combined to generate different final solutions. This can become a very extensive process. If one problem was decomposed into 4 different functions, and 8 solutions were found for each, then a total of 4096 (8x8x8x8) solutions exist for the final product. To explore this solution space systematically and limit the number of possible solutions with methodology, the team can use concept classification trees or combination tables. In a classification tree (Fig. 11a) the problem is divided into different branches. Each branch corresponds to a product function, and these branches will be divided into different subproblems and then into the different solutions found. A good example of this methodology can be found in Sivaloganathan work [7], where he systematically explores the different solutions encountered by engineering students for a specific problem. Combination table (Fig. 11b) is a systematic approach to

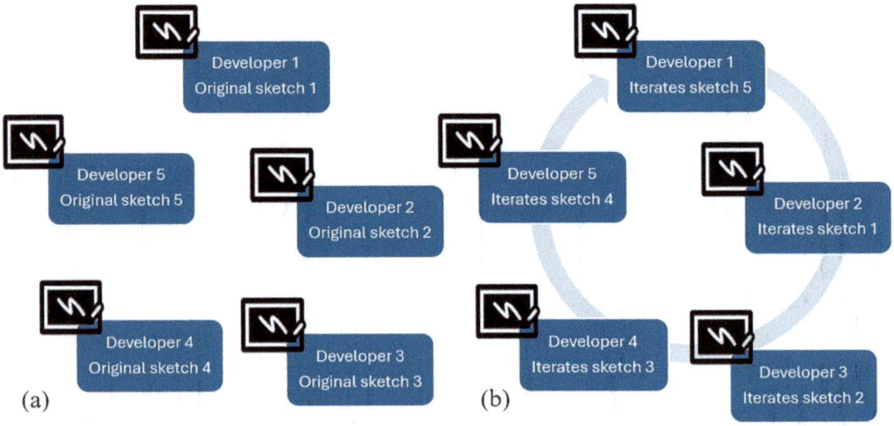

Fig. 10 C-Sketch method, (**a**) initial individual developer sketch creation, (**b**) sketches are circulated between developers and each developer iterates on each other's sketch, and the process is repeated until everyone has contributed to all sketches

represent and explore all the available solutions for a specific problem. Each table column describes all the solutions found for that subproblem. Combining different rows allows the team to discover different solutions, and final concepts arise from this analysis. Eliminating technically impossible solutions and combining coupled subproblems into the same column are two useful ways to reduce the overall number of possible solutions.

In the end, the team should have a wide range of solutions for the product and may choose to create more detailed sketches, virtual 3D models, or 3D-printed prototypes of these ideas. The next step is to select the best ones.

2.4 Concept Selection

Now that the team has generated a large number of potential concepts that meet the end-user needs, they must select the ones best suited for further development. This is not a creative process, but rather a systematic step to compare the different solutions against the criteria provided by the customers: the needs. This is also a very important step, as the chosen concept will influence the final product performance and production cost. Using a structured method allows for a product that is focused on the customer's needs, by selecting the product using the same criteria created in Sect. 2.2. It also makes the decision process clear and well-documented.

The process can be divided into a screening and a scoring phase. While screening is a quick survey and comparison of the available solutions against the end-user needs and product metrics, enabling the team to decide and eliminate the less optimal candidates, scoring will use a detailed score system to compare the

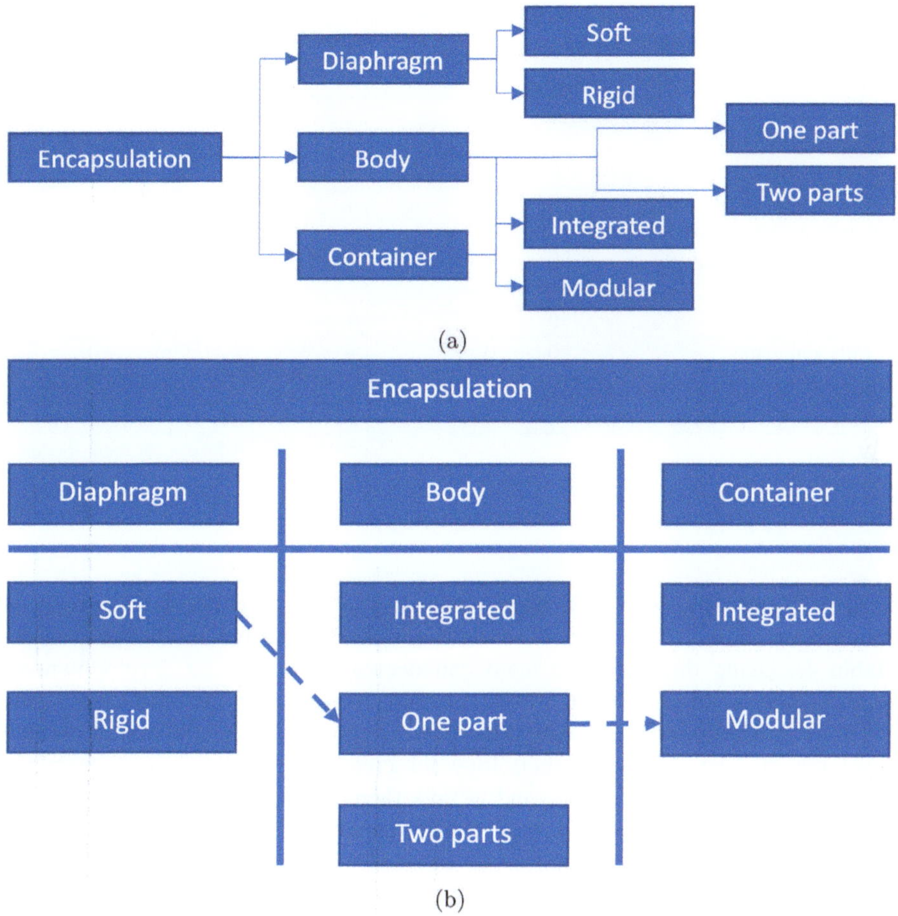

Fig. 11 Examples of (**a**) classification tree and (**b**) combination table, for a digital stethoscope encapsulation

remaining solutions against each other and against competitor products. The base of both stages is to use a screening or scoring matrix. In this matrix, each column represents one of the solutions developed in the previous step, and each line represents one of the developed end-user needs. Concepts will then be eliminated or combined into new ones, taking advantage of the strengths and reducing each other's weak points.

Concept screening was developed by Pugh et al. [26]. Concepts are rated in all categories (end-user needs) against a reference concept. This concept is initially chosen by the team; it can be an existing solution, the market standard, or the gold reference. If none is available, then the team must choose one of the developed concepts as the benchmark solution. Against this solution, all other solutions are

Table 4 Example of a concept screening table, five different concept solutions are screened against the reference solution A. Columns A to E refer to alternative solutions

Need: the lift chair...	A	B	C	D	E
Secures user during sitting/standing motion	0	+	–	0	0
Has arm supports	0	–	–	–	+
Supports user weight during sitting/standing motion	0	–	+	+	–
Describes an anatomically correct motion	0	–	+	0	–
Is fully motorized	0	0	0	0	0
Is fully automatic	0	0	+	+	0
Has a nice design	0	+	–	+	0
# of +		2	3	3	1
# of –		3	3	1	2
Balance		– 1	0	2	–1
Action		Eliminate	Combine	Consider	Eliminate

rated as better (+), equal (0), or worse (–). The final rank of each concept is obtained by adding the number of + signs and subtracting the number of ... signs (Table 4). Using this data, the team can decide to pursue the development of better concepts, eliminate the worst concepts, or combine them to create new concepts. Augustine et al. consider concept convergence an essential part of product development. Therefore, a methodology for creating hybrid solutions, from the initial lot of generated concepts, and to rank them against reference products was developed, with the goal of improving the final product [27].

The goal of concept scoring is to increase the resolution of the decision-making criteria, allowing the team to choose the best possible concept for further development. The previously screened concepts are now scored using a specific scale (1–9, where 1 is the worst and 9 the best) in regard to all the user needs. This is now an absolute scale, and the concepts are no longer rated against a reference solution. The final concept j score CS_j can be calculated using needs weight factors, using Eq. 1:

$$CS_j = \sum w_i \cdot ns_{ij} \tag{1}$$

where w_i is the user need i score weight attributed by the team and ns_{ij} is the score attributed to concept j for need i. Concept scoring is a quantitative method that allows the team to choose the best concept. Still, one must not be blinded by numbers, and authors like Pugh have criticized concept scoring for reducing creativity in product development [26].

2.5 Concept Prototyping and Testing

As mentioned, final product specifications can only be set after some initial prototyping and testing. This stage is crucial to define production processes and to define the materials that will be used in product manufacturing. In order to better understand the implications of the choices, developers can start by prototyping the selected concepts and testing these prototypes in conjunction with future product users. Nowadays, 3D printing technologies play a major role in this stage. 3D printing processes or additive layer manufacturing technologies use material deposition in the form of particles, droplets, or extruded filament to manufacture everything from organs to very large and complex health equipment parts [28]. Flexibility and customizability are two of the main characteristics of 3D printing, enabling it to be used to produce medical devices on a large scale [29]. But these manufacturing techniques are also very useful as prototyping techniques, enabling the quick and inexpensive concept models for testing (Fig. 12) [17].

Testing different medical device concepts is of extreme importance, especially when aiming to develop safer products for health professionals and patients. According to ISO 9241-11 usability is "the extent to which a product can be used by specified users to achieve specified goals with effectiveness, efficiency, and satisfaction, in a specific context of use." Schmettow et al. have developed an extended protocol for usability validation of medical devices. Their goal is to test and benchmark different product concepts by considering several performance measurements as outcome variables. This methodology should include performance capture of different medical tasks after the appropriate training period, while different health professionals should perform the same procedures. Therefore, the random effects of different participants and different tasks will be fully captured. Product concepts are tested regarding human factors, and final product specifications can be set, taking them into account [30].

When testing different concepts, it is also important to assess physicians' motivation to adopt the medical device. Hatz et al. developed the Physicians' Motivation to Adopt (PMA) scale in order to measure this important factor on the decision to adopt and use a new medical device. This scale considers the functional, conformity, power, hedonic, patient benefit, and cognitive dimensions that influence the physicians' decisions. A physician will prefer a functional medical device because it is more reliable, practical, or facilitates procedures. Conformity means the decision is influenced by the organization, peers' advice, or the majority opinion. Power is related to professional recognition, career advancement, becoming a decision-maker, a pioneer, or having an impact on future earnings. Hedonic dimension is related to passion, emotion, state of the art, innovation or excitement, and fun. Patient benefit takes into account an increased well-being, even if it increases personal effort, time commitment, new routines, and learning. And finally, the cognitive dimension is connected with the physicians' analytical mind, intellect, logical thinking, and medical skills [22]. Testing different concepts and assessing

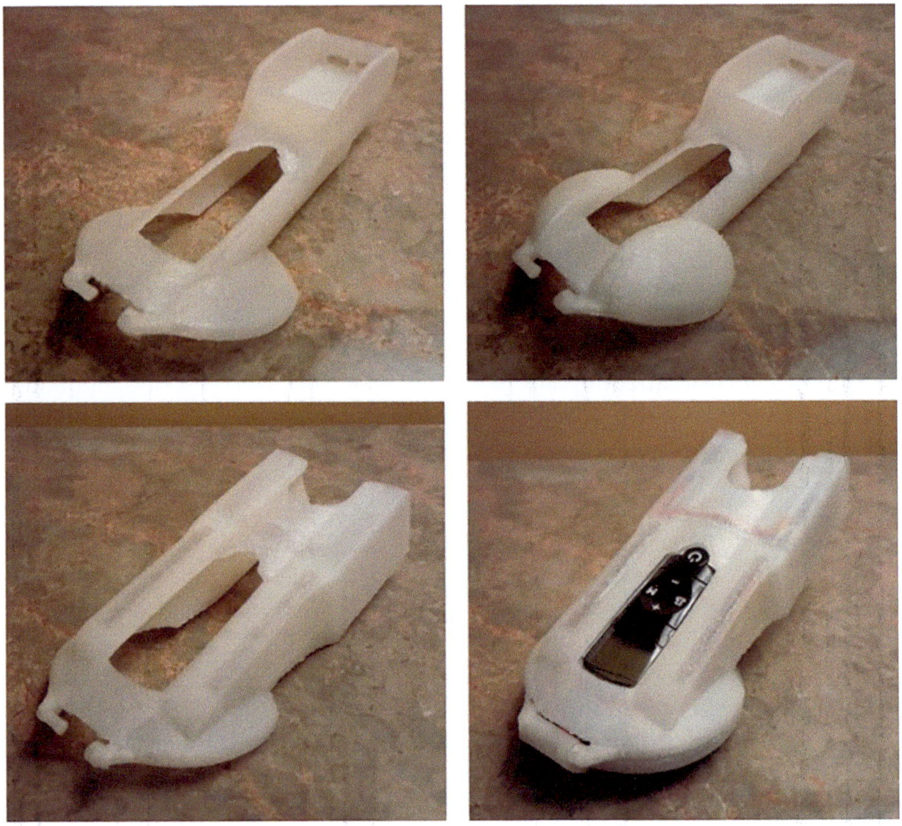

Fig. 12 Example of 3D-printed concept models for a digital stethoscope encapsulation, adding additional features to this medical device

them against the PMA scale is one possible way to rate different concepts and choose the best one, when considering the physicians' opinion.

3 Application to Biomedical Devices

Medical devices are biomedical apparatus with different applications, but with a common goal. This goal is to increase patient health-related quality of life. Medical devices can be used to monitor patient health or to improve a nonmetabolic impaired function. Medical devices also help in the daily activities of health professionals by allowing them to perform medical procedures with high efficiency and safety. Therefore, a medical device can be an external biomechanical device or tool, an implantable mechanical device, an electromedical device, or a diagnostic device [31].

Medical devices are subjected to national and international regulations. A set of standards or norms that medical device manufacturers must comply with in order to develop their products. These regulations set the responsibilities of the different parties involved in the product development process. They are a set of rules or technical requirements that allow for medical device certification, and without this certification, the final product will not be cleared for clinical use. Product developers must include these rules as product specifications at the beginning of the development process. They will define the correct product characteristics, allowing the medical device to be compliant with the corresponding regulations. Many national and international regulations also include testing methods and acceptability criteria. Therefore, these regulations must be part of the product development process from the start [31].

Different countries may have different sets of medical device regulations. Complying with different regulations can increase the duration and cost of the development process, increasing the medical device final cost. This can be considered an obstacle to the use of some medical devices in poor countries and an obstacle to innovation in new products. The International Medical Device Regulators Forum (http://www.imdrf.org/) was created in 2011 with the goal of incentivizing global medical device regulation harmonization [31].

3.1 Medical Device Certification in Europe

The certification process for medical devices within the European Union (EU) included compliance with three different European Directives. The Active Implantable Medical Device Directive (AIMDD 90/385/EEC), the Medical Device Directive (MDD 93/42/EEC), and the In Vitro Diagnostic Medical Device Directive (IVDMDD 98/79/EC) [32]. These regulatory directives will be replaced with the (EU) 2017/745 Regulation on medical devices, by May 2020 (regarding medical devices) or May 2022 (regarding in vitro diagnostic medical devices). (EU) 2017/745 was developed in order to increase patient and health professional safety, using stricter medical device specifications and increasing the risk classification of certain medical devices [31].

In the EU, medical devices are classified in four different classes, accordingly to their risk level. The certification process will also be different for different medical device classes. Class I medical devices have the lowest-risk level classification. They represent general-purpose medical devices. They can be sterile or non-sterile, with or without measuring abilities. Hospital beds, protective gear, sphygmomanometers, or reading glasses are examples of noninvasive devices, without body interaction [32].

Class IIa medical devices have a medium-risk level, as they interact with the human body, for example, exchanging energy with it. Class IIa medical devices may have therapeutic, monitoring, or diagnostic functions. They will generally perform invasive techniques, limited to the body's natural orifices and without hazard potential to the patient. Examples of Class IIa medical devices are thermometers, weighing scales, external hearing aids, powered wheelchairs, or ultrasonic diagnostic health equipment [32].

Class IIb medical devices, on the other hand, perform invasive techniques with hazard potential for the patient. They may be partially or totally implantable in the body and may modify the composition of body fluids. Infusion pumps, surgical instruments, or ventilators are some examples of Class IIb medical devices [32]. Finally, Class III medical devices are classified as a high-risk product. Class III medical devices are responsible for supporting or maintaining human life or preventing health impairment. These devices may connect directly with the central circulatory or nervous systems and may contain drugs or medicinal products. For this type of medical device, a process of scientific review is required for safety and benefit assurance. Neurological and vascular implants, heart valves, or breast implants, are just a few examples of Class III medical devices [32].

Within the EU, medical device manufacturers may choose one of the many independent commercial organizations authorized to certify and attribute the CE mark. These are called Notified Bodies (NB) and act as auditing institutions for each of the medical device classes. After issuing the CE mark, the product is considered compliant with the medical device directives and other safety regulations and may be finally marketed. Accordingly to MDD 93/42/EEC: "Devices, other than devices which are custom-made or intended for clinical investigations, considered to meet the essential requirements referred to in Article 3, must bear the CE marking of conformity when they are placed on the market."

A normal certification process includes (a) check which directives and corresponding annexes will apply to the medical device, (b) choose the conformity route, (c) prepare the product dossier, (d) prepare the technical documentation, (e) prepare the declaration of conformity, and (f) submit the process to the NB [32]. Some medical devices may also require permanent post-certification control and monitoring by the NB [31]. All medical devices must be evaluated regarding their clinical efficacy through a preclinical evaluation [32].

The certification process, or conformity assessment procedure, differs accordingly to the medical device class. But for all cases, compliance is verified by the declaration of conformity, a written statement by the manufacturer that the product meets the applicable medical device directive. For Class I medical devices, the route is very simple and straightforward, requiring only the self-certification described in Annex VII Module A (Fig. 13).

For Class IIa manufacturer must also ensure that the medical device is compliant with part Annex II (quality assurance)) requirements. Finally, the declaration of conformity (Annex VII) must be backed up with the assessment by an NB of verification of conformity (Annex IV), production quality assurance (Annex V), or product quality assurance (Annex VI) (Fig. 14). Class IIb medical devices may be certified accordingly to the full quality assurance route (full Annex II), including the assessment of the technical documentation of at least one sample for each device group (Fig. 14).

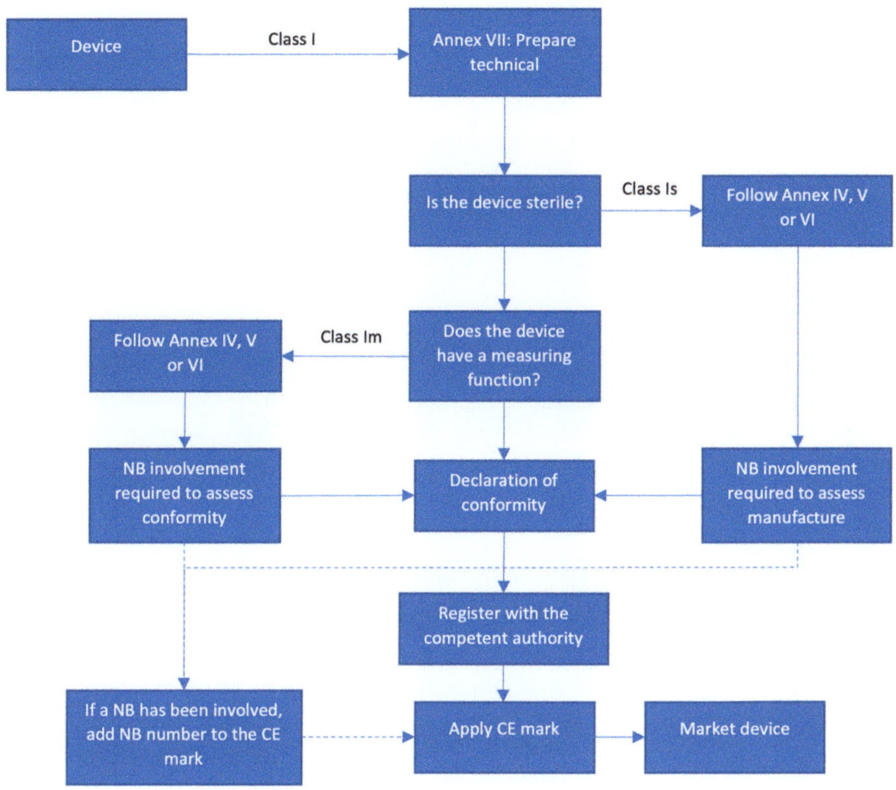

Fig. 13 Class I medical device possible CE marking routes

Class III medical devices are certified as Class IIb (full Annex II) and require permanent monitoring during lifetime by specialized institutions (Fig. 15) [32].

Some medical devices may not require CE marking. As exceptions to Directive 93/42/EEC it is possible to find [32]:

- Custom-made and experimental devices
- For the purpose of investigation devices
- Urgent or unusual circumstances, manufactured devices
- Humanitarian purpose devices
- For in-house use only devices

Some of these medical devices are developed under open-source conditions. Mainly developed by higher education institutions or dedicated biomedical companies, these devices are developed under collaborative strategies. The goal of open-source development is to share the concepts, designs, and technical drawings, improving the performance and safety of these medical devices. As their performance is peer-reviewed, many double checks are in place, assuring the clinical

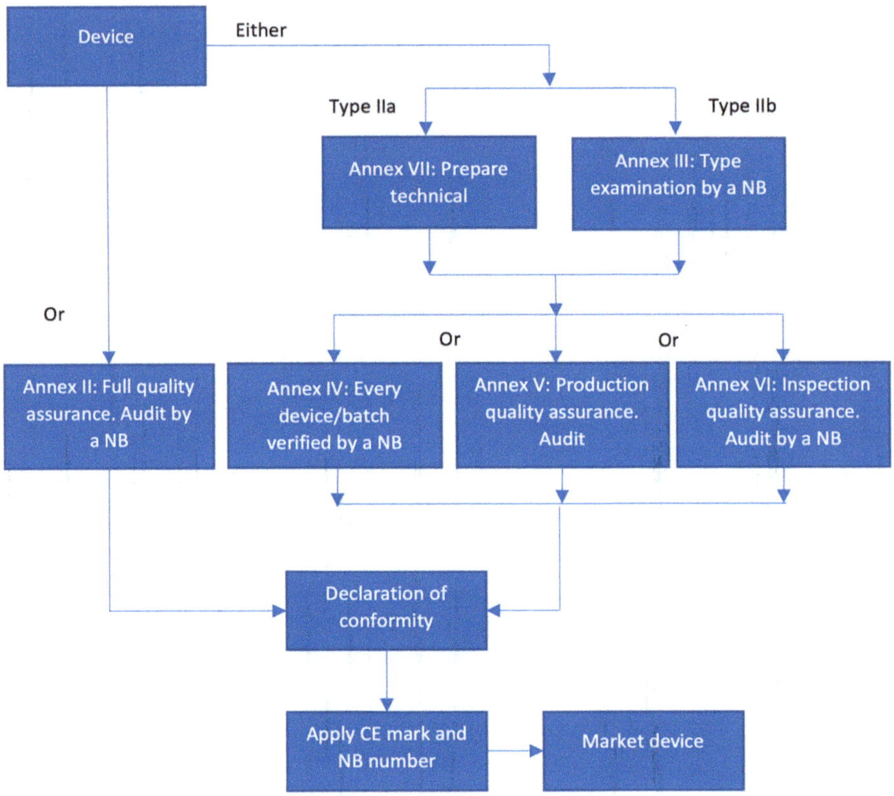

Fig. 14 Class IIa or IIb medical device possible CE marking routes

viability and safety of these products, before the obligatory submission to NB, and before bringing the product to market. This development process also enables a lower development cost, resulting in more sustainable products, to be introduced in developing countries. Additive manufacturing technologies play a major role in the production of this type of medical device, allowing for a more cost-effective means of production and rapid iteration [31]. The Euro-African Open Biomedical Engineering e-platform for Innovation through Education (UBORA) (http://ubora-biomedical.org/) is just one example of an open-source development platform to encourage collaborative design of safer and more efficient medical devices. The UBORA major goal is to solve the current and future healthcare challenges, exploring rapid prototyping, idea sharing, and networking.

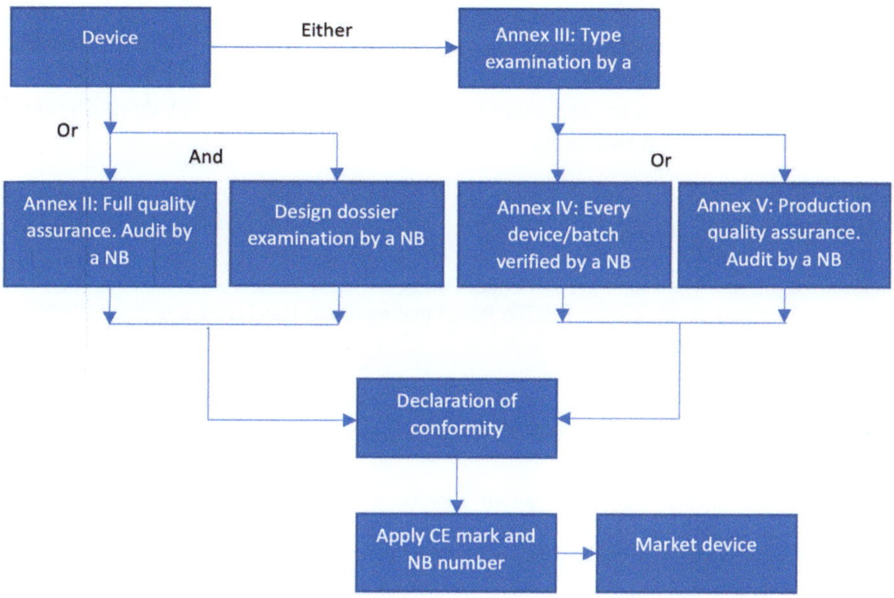

Fig. 15 Class III medical device possible CE marking routes

4 Summary

As with many other types of products, medical devices should always be developed using structured methodologies. These methodologies are a driving factor for innovation, creativity, and safety, allowing the development of user-focused products. As mentioned in this chapter, structured methodologies do not have to be applied in a strict fashion, as development teams can adapt them as needed. But in the end, they provide a fully documented process, with justification for every decision made by the team, and from which it is possible to learn and improve. Still, several key aspects should always be kept in mind when starting the development process of a medical device:

- The process always starts with the identification of an end-user or developer necessity.
- The end user, which in this case can be a patient, a caregiver, or a health professional, should be involved from the start of the development process.
- Development teams should be multidisciplinary and should be able to see or experiment with product functions firsthand.
- This experience and the close relationship with end users are an important source of new ideas.

- Concept generation is an open and creative task; many new and innovative ideas should the generated; and even if they do not appear to solve the problem immediately, they can be iterated and transformed into a working solution.
- Testing these ideas is an important part of concept selection and should be done in collaboration with end users.

The final goal is to develop a successful medical device, focused on and shaped by the end user.

Acknowledgments The author acknowledges Fundação para a Ciência e a Tecnologia (FCT) for its financial support via the project LAETA Base Funding (DOI: 10.54499/UIDB/50022/2020).

References

1. S. Sharples, Medical device design in context: a model of user-device interaction and consequences. Displays **33**, 221–232 (2012)
2. N. Bitterman, Design of medical devices-a home perspective. Eur. J. Intern. Med. **22**, 39–42 (2011)
3. K. Ulrich, S. Eppinger, *Product Design and Development* (McGraw-Hill, New York, 2000)
4. A. Brattström, H. Löfsten, A. Richtnér, Creativity, trust and systematic processes in product development. Res. Policy **41**, 743–755 (2012)
5. W. Durfee, P. Iaizzo, The medical device innovation process. *Engineering in Medicine* (Elsevier, Amsterdam, 2019)
6. Ulrich, The role of product architecture in the manufacturing firm. Res. Policy **24**, 419–440 (1995)
7. S. Sivaloganathan, T. Shahin, M. Cross, M. Lawrence, A hybrid systematic and conventional approach for the design and development of a product: a case study. Des. Stud. **21**, 59–74 (2000)
8. J. Hirtz, R. Stone, D. McAdams, S. Szykman, K. Wood, A functional basis for engineering design: reconciling and evolving previous efforts. Res. Eng. Des. Theory, Appl. Concurr. Eng. **13**, 65–82 (2002)
9. T. Hagedorn, I. Grosse, S. Krishnamurty, A concept ideation framework for medical device design. J. Biomed. Inform. **55**, 218–230 (2015)
10. C. Vincent, Y. Li, A. Blandford, Integration of human factors and ergonomics during medical device design and development: it's all about communication. Appl. Ergon. **45**, 413–419 (2014)
11. J. Martin, B. Norris, E. Murphy, J. Crowe, Medical device development: the challenge for ergonomics. Appl. Ergon. **39**, 271–283 (2008)
12. M. Privitera, M. Evans, D. Southee, Human factors in the design of medical devices—approaches to meeting international standards in the European Union and USA. Appl. Ergon. **59**, 251–263 (2017)
13. A. Stronge, W. Rogers, A. Fisk, Human factors considerations in implementing telemedicine systems to accommodate older adults. J. Telemed. Telecare **13**, 1–3 (2007)
14. J. Rubin, D. Chisnell, *Handbook of Usability Testing: How to Plan, Design and Conduct Effective Tests* (Wiley, London, 2008)
15. R. Branaghan, Human factors in medical device design: Methods, principles, and guidelines. Crit. Care Nurs. Clin. North Am. **30**, 225–236 (2018)
16. A. Gurses, A. Ozok, P. Pronovost, Time to accelerate integration of human factors and ergonomics in patient safety. BMJ Qual. Saf. **21**, 347–351 (2012)

17. B. Sack, Pediatric medical device development by surgeons via capstone engineering design programs. J. Pediatr. Surg. **53**, 493–498 (2018)
18. B. Ranger, A. Mantzavinou, Design thinking in development engineering education: a case study on creating prosthetic and assistive technologies for the developing world. Dev. Eng. **3**, 166–174 (2018)
19. P. Yock, The process of innovating medical technologies-biodesign-textbook (2015)
20. P. Ogrodnik, *Medical Device Design: Innovation from Concept to Market* (Academic Press, New York, 2013)
21. C. Govers, What and how about quality function deployment (QFD). Int. J. Prod. Econ. **46–47**, 575–585 (1996)
22. M. Hatz, T. Sonnenschein, C. Blankart, The PMA scale: a measure of physicians' motivation to adopt medical devices. Value Heal. **20**, 533–541 (2017)
23. G. Pahl, W. Beitz, J. Feldhusen, K. Grote, *Engineering Design: A Systematic Approach* (University of Liverpool Library, 2007)
24. B. Rohrbach, Creative by rules—method 635, a new technique for solving problems. Absatzwirtschaft **12**, 73–75 (1969)
25. J. Shah, N. Vargas-Hernandez, J. Summers, S. Kulkarni, Collaborative sketching (c-sketch)—an idea generation technique for engineering design. J. Creat. Behav. **35**, 168–198 (2001)
26. S. Pugh, *Total Design Integrated Methods for Successful Product Engineering* (Addison-Wesley, Reading, 1991)
27. M. Augustine, O. Yadav, R. Jain, A. Rathore, Concept convergence process: A framework for improving product concepts. Comput. Ind. Eng. **59**, 367–377 (2010)
28. L. Murr, Frontiers of 3d printing/additive manufacturing: from human organs to aircraft fabrication. J. Mater. Sci. Technol. **32**, 987–995 (2016)
29. G. Avventuroso, M. Silvestri, E. Frazzon, Additive manufacturing plant for large scale production of medical devices: a simulation study. IFAC-PapersOnLine **51**, 1442–1447 (2018)
30. M. Schmettow, R. Schnittker, J. Schraagen, An extended protocol for usability validation of medical devices. Res. Design Ref. Model. J. Biomed. Inform. **69**, 99–114 (2017)
31. C. Maria, Safe innovation: on medical device legislation in Europe and Africa. Heal. Policy Technol. **7**, 156–165 (2018)
32. E. French-Mowat, J. Burnett, How are medical devices regulated in the European Union? J. R. Soc. Med. **105**(Suppl), 22–28 (2012)

ScientISST CORE: A Novel Hardware Platform for Rapid Prototyping in Biomedical Engineering

Leonor Pereira, Frederico de Almeida Santos, and Hugo Plácido da Silva

1 Introduction

In recent decades, the emergence of open-source software and low-cost hardware has had a significant impact on a number of areas, including electrical engineering, computer science, and telematics. Notably, the Do-it-Yourself (DiY) hardware community has grown as a result of the reduction of the cost barrier to entry and the shortening of learning curves. Consequently, this has encouraged the development of novel interactive systems that can sense, process, react, and interface the analog and digital worlds, in what is generally referred to as physical computing [1].

Physiological data acquisition has followed a similar path; the use of biosignals in the healthcare domain for diagnostic purposes can be dated back to the beginning of the nineteenth century [2]. But nowadays, due to the significant technological, methodological, and scientific advancements made as a result of the application of biomedical engineering principles and devices beyond the medical domain [3, 4], biosignals are growing in popularity as a research topic within the global scientific community. In fact, the potential uses of biosignals now go far beyond their original applications in the traditional medical domain and into the more expansive and developing field of physiological computing, which includes sports performance [5, 6], performing arts [7], biosignal-controlled human-computer interfaces [8], or biometric applications [9, 10]. Additionally, many of these new applications include the use of Artificial Intelligence (AI) models, another popular topic of research these

L. Pereira (✉) · H. P. da Silva
Instituto Superior Técnico, Universidade de Lisboa, Lisbon, Portugal

IT–Instituto de Telecomunicações, Lisbon, Portugal
e-mail: leonor.pereira@tecnico.ulisboa.pt

F. de A. Santos
Instituto Superior Técnico, Universidade de Lisboa, Lisbon, Portugal

H. P. da Silva et al. (eds.), *Open Source Biomedical Engineering*,
https://doi.org/10.1007/978-3-032-03655-1_2

days, which requires the collection of large volumes of data to increase the results' accuracy and reliability [11].

Unlike physical computing, the sole focus of physiological computing is on the use of biosignals, which brings its own particular difficulties given the more complicated requirements associated with acquiring physiological data. These include the demand for higher Signal-to-Noise Ratio (SNR), increased sampling rate accuracy, and improved signal resolution. Consequently, these requirements make it so that, generally speaking, even the most basic biosignal acquisition tasks require prohibitively expensive hardware (typically well above ~US$10 K).

To address these challenges, more recently, efforts have been developed to create enabling low-cost instruments for physiological sensing, similar to the DiY electronic prototyping platforms used in physical computing projects (e.g. Arduino and Raspberry Pi), which have been widely adopted [12]. Nonetheless, a number of limitations still affect the resources available in this field, namely, limited functionality (e.g. devices that can only measure one or two pre-defined biosignal modalities), limited applicability due to lack of access to raw data (e.g. only the post-processed cardiovascular signals are provided by most heart rate monitors), preclusion of important operations for educational and experimental purposes (e.g. direct control/interaction with the hardware) due to industrial property issues, or technical limitations for certain biosignal acquisition tasks.

In this work, we present ScientISST CORE,[1] a reference hardware design intended to expedite experimentation in biomedical engineering. With ScientISST CORE, we aim to further democratize rapid prototyping of applications which involve biosignal acquisition and processing, while simultaneously addressing certain drawbacks of comparable low-cost platforms. Our main objective is to offer a high-performance tool for developing imaginative and inventive engineering projects to the physiological computing community. Furthermore, owing to the ecosystem of tools built around it, which incorporates sensors and several programming APIs, it can be immediately utilized in an out of the box manner. In the long term, we hope that this tool can aid in mitigating the misconception that many students, scholars, and practitioners have of biosignals as a complex and hard-to-reach topic due to lack of experience in this area.

The rest of the chapter is organized as follows: Sect. 2 presents related work; Sect. 3 summarizes the approach taken in developing the ScientISST CORE base hardware (for a more in-depth characterization see CITE); Sect. 4 presents the methodology used in the quantitative tests performed to characterise the temporal uncertainty (in terms of sampling rate accuracy) when using Bluetooth or Wi-Fi for data transfer, or when storing the data in a memory card; Sect. 5 describes the performance evaluation of the hardware in these tests; and, finally, Sect. 6 outlines the main conclusions.

[1] https://scientisst.com

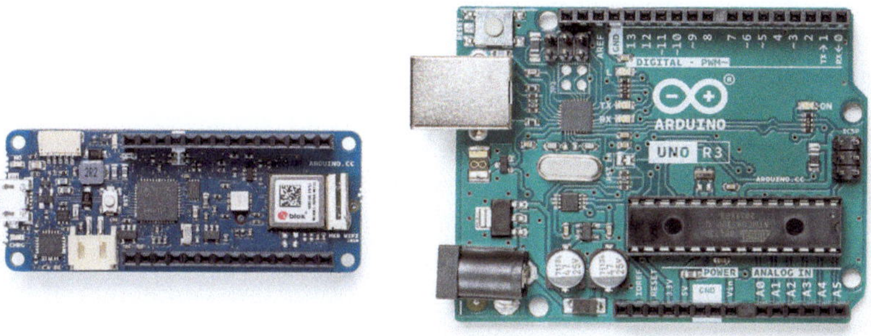

(a) (b)

Fig. 1 Example of two popular boards from the Arduino ecosystem. (**a**) Arduino MKR WiFi 1010. (**b**) Arduino UNO

2 Related Work

Created in 2005 by Massimo Banzi and his team, the most renowned open-source electronics platform is by far the Arduino [13]. This is due to the several advantages it offers over other systems, such as: a lower cost; simple and clear programming environment (Arduino IDE) that runs on Windows, Mac OS, and Linux operating systems; and extensible open-source software and hardware. Over the years, numerous Arduino boards have been developed with distinct technical specifications. These can have different microcontroller chips, operating voltages (usually 3.3 or 5 V), number of analog and digital pins, Analog-to-Digital Converter (ADC) resolution (most commonly 10 bits), connectivity modules (e.g. Wi-Fi, Bluetooth, BLE), and communication protocols available (e.g. UART, Inter-Integrated Circuit (I2C), Serial Peripheral Interface (SPI)).

A recent example is the Arduino MKR WiFi 1010[2] (Fig. 1a), which includes a Microcontroller Unit (MCU) running at a clock speed up to 48 MHz, providing 8 digital I/O pins, 7 analog inputs with 12-bit resolution, 1 analog output, and 32 KB of Static Random Access Memory (SRAM). It can be powered by an external battery and has Wi-Fi, Bluetooth (BT), and Bluetooth Low Energy (BLE) connectivity, making it particularly suitable for Internet of Things (IoT) applications. The most well-known version, albeit with simpler technical specifications, is arguably the Arduino UNO[3] (Fig. 1b), which has a 16-MHz maximum MCU clock speed, 10-bit ADC resolution, 2 KB of SRAM, and no Bluetooth or Wi-Fi connectivity.

However, most Arduino-based toolkits are more geared towards physical computing applications [14, 15] and are not specifically designed with support for biosignal acquisition in mind. As a way to circumvent this issue, Arduino-

[2] https://docs.arduino.cc/hardware/mkr-wifi-1010

[3] https://docs.arduino.cc/hardware/uno-rev3

compatible biosignal acquisition sensors have been developed over the years, as is the case of the Pulse Sensor[4]—a plug-and-play Photoplethysmography (PPG) sensor for heart rate detection—or the MyoWare Muscle Sensor[5] for Electromyography (EMG) data acquisition.

Although these sensors can be useful as a starting point in an educational setting, Arduino-based approaches cannot meet certain requirements for biosignal acquisition that make it unsuitable for physiological computing applications. These stringent requirements include a high sampling rate accuracy and noise rejection capabilities, for which previous studies have shown that Arduino platforms fail to meet due to considerable drift, skew, and jitter in the sampling rate, combined with high sensitivity to noise, harmonic distortion, and crosstalk between channels [16, 17].

The OpenEEG Project [18] is one of the most comprehensive efforts made to date both in terms of biosignal acquisition and processing. Over the years, numerous people in a community-driven effort have contributed to this project, which aims to develop affordable hardware and software tools that can be used by any individual interested in exploring Electroencephalography (EEG). Their Modular EEG device, which can be assembled by the user or bought already pre-assembled, is made up of two or more EEG amplifiers and a 6-channel signal acquisition board that connects to a PC via a standard serial cable, where a number of open-source apps (e.g. BioEra) can be used for biofeedback and EEG analysis. There are other commercially available boards for EEG data acquisition, such as the OpenBCI Cyton board, an Arduino-compatible, 8-channel bio-amplifier with a 32-bit processor and its own purpose-built firmware, that works with the OpenBCI GUI to allow a user-friendly interaction with the board. Although this is a low-cost device, studies have shown it has a good classification performance in motor imagery detection [19]. However, these boards are primarily focused on a single biosignal modality, EEG data, hence using relatively low sampling rates and having bulky form factors.

In 2017, Libelium launched MySignals[6] (Fig. 2), one of the most complete multimodal development platforms for medical devices and e-Health applications, which includes 16 sensors capable of measuring more than 20 biosignals, such as pulse, blood oxygenation (SpO2), Electrocardiogram (ECG) signals, breath rate, blood pressure, glucose levels, muscle EMG signals, glucose levels, and electrodermal activity (EDA), among others. There are two versions of this platform available: the Software Development Platform; and the Hardware Development Platform.

The MySignals software includes a high-level firmware capable of controlling all the sensors from the device simultaneously, storing the acquired data and derived features on the device to display them on a TFT LCD, or uploading them to the cloud

[4] https://pulsesensor.com/

[5] http://www.advancertechnologies.com/p/myoware.html

[6] http://www.my-signals.com/

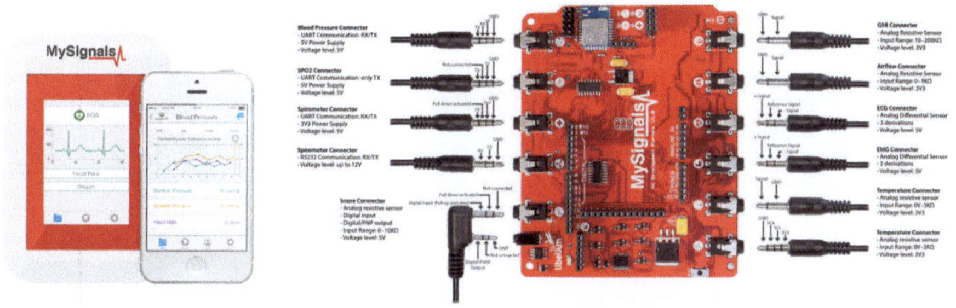

(a) (b)

Fig. 2 MySignals development platforms. (**a**) Software. (**b**) Hardware

(using Wi-Fi) or a mobile phone (using BLE) to instead display them on a web server application or mobile app, respectively. On the other hand, the MySignals hardware is Arduino-compatible, meaning it can be programmed through the Arduino IDE. For this purpose, it includes high-level libraries to easily manage all the sensors integrated in the board, offering a greater degree of customizability. However, the processor is limited, so it is not possible to manage all sensors and have wireless communication and other features simultaneously.

The main advantage of the MySignals platform is the fact it integrates a wide variety of sensors with embedded processing algorithms to remove noise and other artefacts from the acquired data, as well as extract high-level features. However, some drawbacks can still be identified in this platform, mainly regarding how the pre-processed data from some of these sensors is not made available (limiting their applicability), lack of information on the processing algorithms used, and the platform's poor flexibility for custom hardware configurations (on account of the fact all sensors are provided as a single unit). Furthermore, both of the MySignals development platforms were discontinued in 2020, making it impossible to obtain them past this point.

Today, perhaps the most comparable tool to the Arduino for applications involving biosignal acquisition is the BITalino [20]. Like the Arduino, it is a low-cost hardware device, but its design was originally created by our group and geared towards biosignal acquisition, by building on the guiding principles of existing physical computing hardware platforms. Several versions have been developed, but the most complete would be the BITalino (r)evolution, an all-in-one hardware device, designed to meet the learning and prototyping needs of any user interested in experimenting with biosignal acquisition, having even been recognized by the European Commission's Directorate General for Communications Networks, Content, and Technology (commonly known as DG-CONNECT) as the best "Industrial and Enabling Tech" of 2017 within the scope of the organization's Innovation Radar Prize.

In order to maximize versatility, BITalino is composed by a set of modular blocks. The analog front-end incorporates several individual sensors for bioelec-

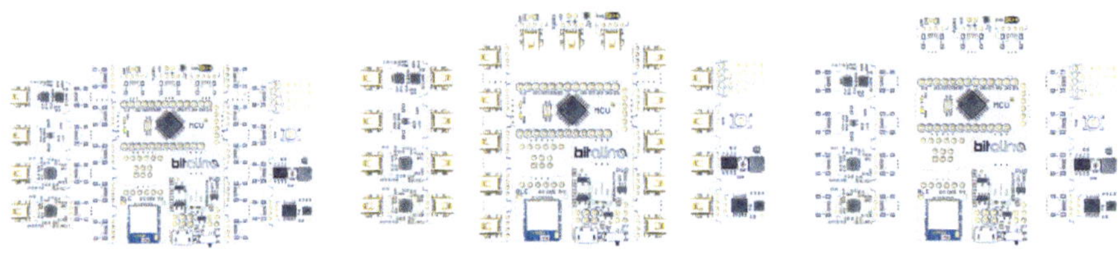

(a) (b) (c)

Fig. 3 Different configurations of the BITalino biosignal acquisition hardware. (**a**) Board. (**b**) Plugged. (**c**) Freestyle

trical and biomechanical data acquisition, including ECG, EEG, EMG, EDA, and ACC, as well as a light sensor (LUX). BITalino is also equipped with actuators, namely, a LED analog block, to enable synchronization with third-party equipment (e.g. a computer screen or video camera), a buzzer for audio feedback, and a Digital-to-Analog Converter (DAC). There is also a block with a push button for manual event annotation, as well as a prototyping board block that serves as an interface for custom accessories. By default, the BITalino platform comes as a single board where the analog and digital ports on the control block are already connected to the on-board sensors. However, this system was designed in a way that enables the physical detachment of each individual block from the main board, allowing for its use in one of the three configurations depicted in Fig. 3.

Furthermore, the architecture is supported by a control block based on the 8-bit Atmel AVR-RISC microcontroller (ATmega328P), a power management block for circuit voltage regulation and LiPo battery charging, and a Bluetooth or BLE/BT dual-mode communication block for wireless data transfer. For physiological data acquisition, BITalino has a purpose-built firmware, optimized to acquire a maximum of six analog and four digital channels at up to 1000-Hz sampling rate with maximum performance. To communicate with BITalino and access sensor data, users can opt from a number of available programming APIs or software applications (e.g. OpenSignals).

Over the years since its development, numerous studies have demonstrated the reliability of the BITalino in applications involving biosignal acquisition. In a study conducted by Batista et al. [21], the quality of data acquired by the BITalino (r)evolution and an older version of this system was compared against a reference device, and it was found that the signals obtained from the BITalino platforms were well correlated with the latter device. Furthermore, in a study conducted by Guerreiro et al. [17], using the aforementioned older version of the BITalino, results showed that even this device outperformed Arduino platforms in terms of robustness to noise and sampling rate accuracy, as result of the latter's high skew and jitter caused by the considerable runtime overhead introduced by the Arduino bootloader and libraries.

Even though these findings demonstrate that the BITalino is better suited for biosignal acquisition, this device still carries with it some limitations. Namely, the BITalino platform cannot support certain physiological computing applications

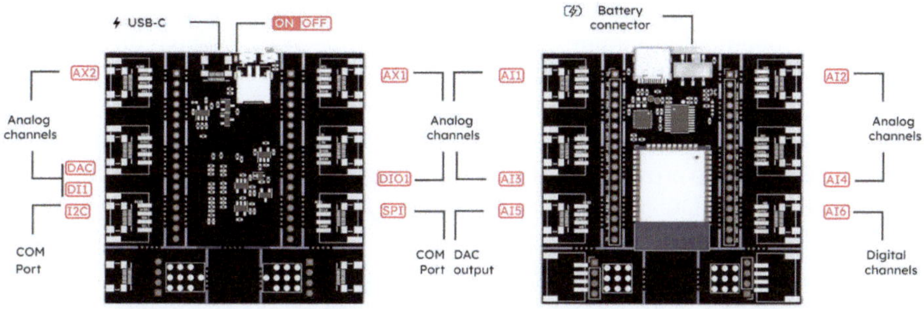

Fig. 4 ScientISST CORE development board

due to its technical specifications, such as the maximum sampling rate (e.g. in phonocardiography, acoustic signals require sampling rates of 4000 Hz [22]) and the resolution of the analog input channels (BITalino only has 4 channels with 10-bit resolution and 2 with 6-bit). On top of this, BITalino does not have I2C or SPI ports, meaning it does not support communication with external devices that use these protocols. Furthermore, its use is limited in an educational setting, since it is incompatible with the Arduino IDE, a programming environment first learned by most students.

3 Proposed Approach

The ScientISST CORE (Fig. 4) board is a novel low-cost hardware architecture specifically designed to overcome the limitations described in Sect. 2 through its improved technical specifications for biosignal acquisition. Hereinafter, for the sake of simplicity, this device is referred to as CORE. The CORE board is built upon the ESP32-WROOM-32[7] MCU module. At the core of this module is the ESP32-D0WDQ6[8] chip, which includes two low-power Xtensa® 32-bit LX6 microprocessors, with a configurable CPU clock frequency from 80 MHz to 240 MHz. Additionally, this chip features 520 KB of SRAM and 4 MB of flash memory.

3.1 Analog Interfaces

CORE exposes six main analog input ports to which any sensor with a compatible analog output can be connected using a Grove cable[9] or a micro HDMI cable

[7] https://www.espressif.com/sites/default/files/documentation/esp32-wroom-32_datasheet_en.pdf

[8] https://www.espressif.com/sites/default/files/documentation/esp32_datasheet_en.pdf

[9] https://wiki.seeedstudio.com/Grove_System/#jump9

(depending on the connector soldered to the board). All analog inputs are connected to a signal conditioning circuit designed to bring the input voltage down to the measurement range of the internal ADC from the ESP32 module (ADC1, 0–1.1 V). This ADC has a higher resolution (12-bit) when compared to the state of the art, allowing for more precise measurements. When using our purpose-built firmware, CORE also provides a battery level using one channel from the ESP32 12-bit internal ADC (ADC2), without sacrificing any of the effective data acquisition channels.

Lastly, the CORE board integrates a high-resolution external ADC (24-bit effective resolution) with two additional analog input ports that are more suitable for tasks that require a higher degree of precision in the conversion process. In total, CORE exposes a total of 8 analog inputs.

3.2 Digital Interfaces

SPI and I2C ports are also included in the CORE board, to give it the capacity to establish a communication line with external devices that use these types of digital interfaces and could be useful to support the applications being developed, such as memory card modules for SPI, LCD displays, and various sensors (e.g. distance, temperature, humidity sensors) for I2C.

Additionally, CORE exposes a pair of digital input and output pins (1-bit), useful for receiving synchronization signals from external devices or event annotation switches, as well as for sending synchronization signals to external devices or triggering external interfaces (e.g. a LED). A DAC port is also integrated to provide access to one of the two 8-bit DAC channels from the ESP32 module, which can be used to convert a digital signal into an analog voltage output.

At the bottom of the board, there are four breakout connectors (two Grove and two female micro HDMI) that are not connected to the rest of the board, but are included to facilitate the connection of external accessories, thus increasing its potential customisation.

3.3 Power

CORE is battery operated, which means we can guarantee its independence from any high-voltage power source during normal operation. However, this can affect biosignal measurements, if the board is not properly designed to take into account the voltage drop associated with the normal battery discharge and fluctuations in the power supply introduced by modules with high peak currents, as is the case of the aforementioned connectivity modules and digital integrated circuits [23].

As such, in order to ensure a steady constant supply through all operational conditions and reduce the external supply voltage to the board's operating voltage,

Table 1 Effect of resistor(s) configuration on ScientISST CORE's operation

	Effect on resistor(s) removal		
	Battery charge	Flashing possible	Analog circuitry decoupled from USB-5V
Resistor(s) to be removed			
RVBUS RUSBP RUSBN	Not possible	Not possible	✓
RLIPO	Only in OFF mode	Only in OFF mode	Not completely
RVBD	✓	Only in ON mode if battery is plugged	Not completely
RLIPO RVBD	Only in OFF mode	Not possible	✓
None	✓	✓	Not completely

voltage regulators were incorporated into CORE. The board has two separated regulated 3.3-V outputs and grounds to independently power the analog and digital parts of the acquisition circuit, in order to protect the biosignal analog circuitry from the noise produced by the digital circuitry in the power lines (e.g. when transmitting data). This way, we can achieve maximum analog performance and prevent corruption from digital noise. Additionally, the board also has a regulated 1.65-V output to provide an offset voltage for biosignal sensors that have a bipolar measurement range (e.g. for voltage potential differential sensors used to acquire ECG, EMG, and related signals).

Additionally, CORE is fitted with a USB-C connector, which is used to recharge the battery, among other things detailed in Sect. 3.4. However, when a device is being powered by an external power source through the USB cable, it could be a potential hazard, if sensors that are coupled to a person's body are connected as well. There is an increased risk of electric shock on account of the potentially high common-mode voltage on the USB signals and the low impedance path to the body via electrodes. To protect the user from this risk, several 0Ω resistors may be removed from the board in different combinations, as a way of providing distinct degrees of protection, by decoupling the analog circuitry from the USB power source. However, it is important to note that removing these resistors can limit the ability to reprogram the device or recharge the battery, so there is a trade-off between added protection to the user and some functionalities, as summarized in Table 1.

3.4 Communication

Apart from recharging the battery, the USB-C connector on the CORE board makes it possible to establish a serial communication line between the board and a PC for data transmission and reprogramming purposes.

For wireless data transfer, the CORE board supports three wireless communication protocols: Wi-Fi (using the 802.11 b/g/n wireless standard); Bluetooth v4.2; and Bluetooth Low Energy (BLE). By adopting both Wi-Fi and Bluetooth technology (classic and low energy), in combination with the battery-powered operation, we can ensure electrical isolation of the user from high-voltage power sources, as well as broader interoperability when compared to other systems which only use one of these modules. Alternatively, because the CORE board includes a SPI port, acquired data can be stored in an SD card that is connected to this port using a memory card module. It is important to note that, when acquiring data to be stored to an SD card, only two modes are supported at the moment: all internal analog channels at 1000 Hz sampling rate, or all internal analog channels and 1 or 2 external analog channels at 100 Hz. After data acquisition, the data can then be retrieved by removing the SD Card and using a provided Python script to convert the binary files to CSV files.

3.5 Firmware

By default, the CORE board already has a purpose-built firmware for physiological data acquisition, developed using the native ESP32 development framework, ESP-IDF.[10] This firmware manages the behaviour of the system in terms of, e.g. the sampling rate used and the sampled channels. To control the wireless data streaming over Wi-Fi, BT, or BLE, record and plot the data being received; the firmware supports two software APIs: the recommended ScientISST API (available natively in Python[11]), which has more features and is more actively maintained, and the deprecated BITalino API, which is only supported for backwards compatibility with the BITalino.

To transmit the data from the sampled input channels selected by the user, this data is first formatted into packets of variable size, ranging from 20 bytes (when all analog channels are selected) to 7 bytes (when only one 12-bit analog channel is selected). The data frame includes a 4-bit Cyclic Redundancy Check (CRC) code, which enables frame data integrity check and allows for the detection of data packet inconsistency (e.g. caused by loss of packets in communication). Additionally, the data packet no longer contains a 12-bit packet sequence number to identify cases of packet loss, as seen in CITE [24]; instead, it now includes a 36-bit number that

[10] https://docs.espressif.com/projects/esp-idf/en/latest/esp32/#

[11] https://github.com/scientisst/scientisst-sense-api-python

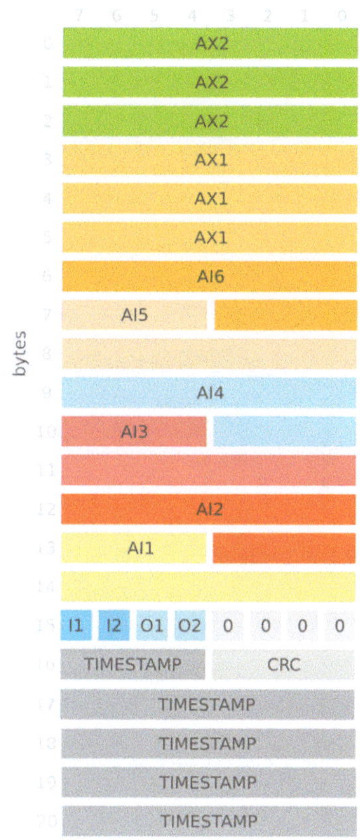

Fig. 5 Data packet structure, namely, timestamp, CRC code, digital input channels (I1 and I2) and output channels (O1 and O2), and analog channels associated with the internal ADC (AI1 to AI6) and to the external ADC (AX1 and AX2)

corresponds to the real timestamp of data acquisition (in microseconds). In order to obtain the timestamp relative to the start of the acquisition, the user simply has to subtract the timestamp from the first packet. With the timestamp, cases of packet of loss can still be identified, and now it is possible to ascertain more sources of these losses, such as the device's internal buffer becoming full, the actual sampling rate being different from the requested sampling rate, and any freezing of the firmware that may occur. Additionally, it is no longer needed to use its estimated value from the selected sampling rate, which has a positive effect on some aspects of performance, as will be shown in Sect. 5. Figure 5 shows the structure of the data packets streamed by the CORE board.

Although there are a number of software APIs available with high-level functions to control the operation of the board, as was mentioned before, the configurable settings can also be changed by sending 1, 2, or 3 byte commands from the base station directly to the CORE board. Figure 6 shows the set of commands recognised by the purpose-built firmware.

Regarding the operation commands, CORE can operate under "Idle Mode", where the device is in standby and waiting to receive further commands; "Live

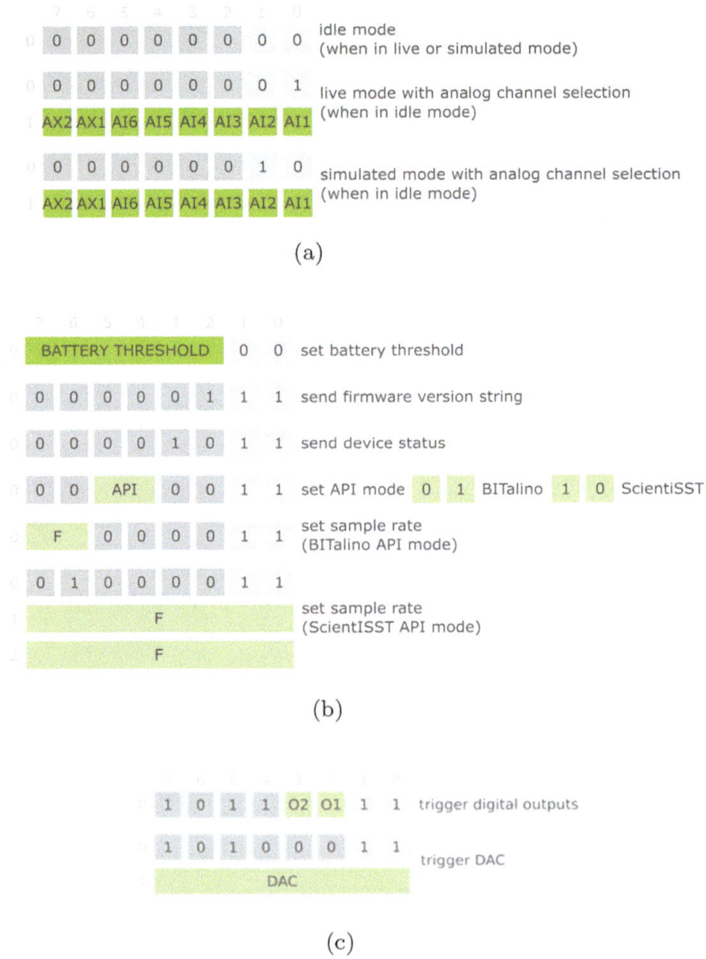

Fig. 6 Commands recognized by the ScientISST CORE firmware. (**a**) Operation commands. (**b**) Configuration commands. (**c**) Trigger commands

Mode", where data is being acquired in real time from the analog channels selected; and "Simulated Mode", similar to the previous mode but in which the transmitted data is synthesised.

On the other hand, the configuration commands allow the user to modify certain settings related to the operation of the device, including the battery threshold, API mode (ScientISST CORE or BITalino), and sampling rate, as well as request information regarding the firmware version and device status. Furthermore, the trigger commands enable the use of the digital outputs and the DAC by sending a simple 1- or 2-byte command, respectively. Access to this low-level control of CORE expands its interoperability, by allowing the user to interface this device with other embedded systems. Additionally, it can be used in academic and pedagogical activities to explore data unpacking, CRC check, and other concepts related with digital communication using data acquired in real time.

Unlike other devices found in the state of the art, CORE is compatible with the Arduino IDE, meaning that custom firmware can be more easily created using this programming environment. Still, the Arduino framework, due to its nature, produces firmware that use more memory and are less performing when compared to the ones developed using the native ESP32 development framework. This negatively impacts the system's sampling rate accuracy, hindering the use of the device in research-grade applications.

Table 2 summarizes some of the features of CORE comparatively to the hardware platforms found in the state of the art, highlighting how the former outperforms the majority of them.

4 Experimental Evaluation Methodology

In a previous study [24], the performance of the CORE board was tested in terms of real-time data acquisition and streaming, using BT, by evaluating the temporal accuracy of the ADC process (i.e. sampling rate accuracy), when acquiring square waves with 10 or 100 Hz frequency at different sampling rates, using a variety of ADC channel configurations. The results showed that the real-time signal acquisition was done reliably and accurately, with no loss of data at sampling rates of up to 1 kHz for all analog channels, up to 2 kHz for all external analog channels, up to 8 kHz for a single external analog channel, and up to 10 kHz for a single internal analog channel.

However, since these tests were performed, the data packet structure has been altered to include the timestamp of data acquisition, as was mentioned in Sect. 3.5, which would theoretically improve usability as the timestamps no longer have to be deduced from the selected sampling rate. To confirm this, the ADC temporal accuracy tests, using BT as the communication mode, were repeated.

Additionally, these same tests were performed again using Wi-Fi to transmit the data packets, as data packet integrity with this communication mode has not been tested before. To do this, the firmware first needs to be configured to switch the communication mode to Wi-Fi, by opening the Espressif IoT Development Framework Configuration menu and navigating to the "Communication configuration" menu, as depicted in Fig. 7a–e. After switching the communication mode, a new "Wi-Fi configuration" menu will be available for the user to provide the Wi-Fi SSID and password, host IP and port, after which the firmware can be flashed onto the board.

Lastly, the ADC temporal accuracy and data packet integrity were tested again, but using an SD card to store the acquired data instead of transmitting it. Again to use this mode, the firmware needs to be configured first and then re-flashed onto the board, as is depicted in Fig. 7a–c and f. For more detailed instructions on how to configure the firmware and flash the board, we refer the reader to CITE [24].

Table 2 Comparison between ScientISST CORE and related work

	Features						
	MCU						
Platforms	Clock [MHz]	Flash memory	SRAM [kB]	Cores	Analog input channels	Data link	Arduino IDE compatibility
Arduino UNO	16	32 kB	2	1	6 (10-bit)	USB + Serial + I2C + SPI	Yes
Arduino MKR WIFI 1010	48	256 kB	16	1	7 (12-bit)	USB + Serial + I2C + SPI + BT + BLE + Wi-Fi	Yes
OpenEEG	16	16 kB	1	1	8 (10-bit)	Serial	No
MySignals HW	16	32 kB	2	1	6 (10-bit)	Serial + Wi-Fi + BLE	Yes
BITalino	8	32 kB	2	1	4 (10-bit) + 2 (6-bit)	BT + BLE + Serial	No
ScientISST CORE	80-240	4 MB	520	2	6 (12-bit) + 2 (24-bit)	USB + Serial + I2C + SPI + BT + BLE + Wi-Fi	Yes

Fig. 7 Espressif IoT development framework configuration menu. (**a**) Top. (**b**) Component config. (**c**) ScientISST configuration. (**d**) Hardware version. (**e**) Communication configuration. (**f**) Save configuration

Similarly to the tests conducted in CITE [24], to characterize the temporal accuracy of the ADC process when transmitting the data using BT or Wi-Fi, square waves with 100-Hz frequency, synthesized by the Tektronix AFG3102 Function Generator, were acquired using the CORE board at different sampling rates. Data acquisition was done using all possible combinations of the 12-bit and 24-bit analog channels from the internal and external ADCs, respectively, up to the maximum sampling rate achievable without data acquisition becoming unreliable, due to errors.

On the other hand, for the SD card, only two modes are supported by the firmware at the moment: all internal analog channels acquiring data at a sampling rate of 1000 Hz and all internal analog channels and 1 or 2 external analog channels at a sampling rate of 100 Hz. As such, data acquisition was also performed with a 100 Hz square wave for the first mode, but with a 10 Hz square wave when using all analog channels due to this limit in the maximum sampling rate. This limit is due to the fact that the SD card and the external ADC use the same SPI communication lines, and, because the former has a higher clock speed than the latter, by lowering the clock speed to the level of the external ADC, writing data to the SD card becomes slower and blocks data acquisition.

For all the tests performed, regardless of communication mode, data was acquired using the CORE board from the selected channels for 240 seconds, corresponding to 24,000 cycles for the 100 Hz square wave and 2400 cycles for the 10 Hz square wave. All synthesized signals were also acquired using an oscilloscope (Tektronix MDO3054[12]), to ensure these signals were properly synthesized.

From the signals acquired by the CORE board, we extracted statistics from the measured wave period values, namely, their minimum, maximum, mean, median, and standard deviation. Additionally, the drift caused by the accumulation of temporal inaccuracies in the ADC process, which lead to errors in the acquisition of biosignals (namely, in the measured signal period), was also characterized. This drift was obtained by computing the value of the difference between the timestamps when a rising edge occurred in the wave acquired by the CORE board comparatively to when this same edge should have occurred in the synthesized signal. Similarly to the wave period measurement values, statistics were extracted from the obtained drift values, namely, their minimum, maximum, mean, median, standard deviation, and cumulative value, which will be discussed in the next section.

5 Results and Discussion

Tables 3 and 4 show the most relevant statistics extracted from the wave period measurements and the drift characterization results for an internal (AI1) and an external analog channel (AX1), respectively. These statistics include the minimum, maximum, and mean of the measured wave period values, along with the measured drift value at the end of acquisition and the compensation per second, which translates to the expected absolute value of drift (in μs) accumulated per second. Note that the values shown in these tables for each possible ADC channel combination correspond to those obtained with the maximum achievable sampling rate for that combination, up until the point data acquisition became compromised, due to errors. To find the results of all statistics mentioned in Sect. 4, for every sampling rate value tested up to the maximum, we direct the reader to Tables 5, 6, 7, and 8, which can be found in Annex 1.

The results displayed in Tables 3 and 4 show that, for the same channel configurations tested in [24], we were able to achieve higher sampling rates without data acquisition becoming compromised (e.g. 16 kHz compared to 10 kHz for one internal channel, 10 kHz compared to 8 kHz for one external channel, 4 kHz compared to 1 kHz for all internal channels, and 2 kHz compared to 1

[12] https://www.tek.com/en/datasheet/mixed-domain-oscilloscopes

Table 3 Waveform period measurement and drift characterization results for the AI1 channel, using different ADC channel configurations

Number of active channels			Measured wave period statistics			Measured drift statistics	
External	Internal	Sampling rate (Hz)	Min (μs)	Max (μs)	Mean (μs)	Cumulative value (μs)	Compensation (μs/s)
0	1	16,000	9747	10,234	10,000	1620	7
	2	10,000	9860	10,130	10,000	1581	7
	3	5000	9794	10,205	10,000	1607	7
	4	5000	9776	10,245	10,000	1603	7
	5	4000	9729	10,260	10,000	1756	7
	6	4000	9737	10,255	10,000	1717	7
1	1	8000	9869	10,147	10,000	1625	7
	2	8000	9844	10,155	10,000	1753	7
	3	5000	9794	10,217	10,000	1607	7
	4	5000	9748	10,232	10,000	1752	7
	5	4000	9720	10,283	10,000	1466	6
	6	2000	9494	10,512	10,000	1507	6
2	1	5000	9769	10,227	10,000	1788	7
	2	4000	9713	10,286	10,000	1749	7
	3	2000	9490	10,507	10,000	1484	6
	4	2000	9484	10,517	10,000	1516	6
	5	2000	9497	10,501	10,000	1496	6
	6	2000	9491	10,513	10,000	1530	6

kHz for all channels). The reason behind this performance improvement is most likely the addition of the real timestamp of data acquisition in the transmitted data packets, since deducing this timestamp from the selected sampling rate adds errors, caused by inaccuracies in the chip's timers, calculation processing times, and/or delays in certain processes that are running simultaneously, which accumulate over time. These add to the temporal inaccuracies in the ADC process inherent to the functioning of the CORE board, leading to higher perceived drift and consequent errors in the acquisition of biosignals, namely, in the measured signal period. By adding the real timestamp, these additional temporal inaccuracies are removed, allowing for reliable data acquisition at higher sampling rates.

Additionally, as was observed in [24], the maximum sampling rate achieved for each channel configuration tends to decrease as the data packet size increases (i.e. as more channels are selected), with the highest value registered being 16 kHz for only one internal channel (8-byte data packet) compared to only 2 kHz for all channels (20-byte data packet). This is to be expected, since the CORE board is only able to transfer a limited amount of data in a reliable manner to a base station in a given amount of time. This property, called throughput, is limited by both the number of ADC channels being sampled and the sampling rate used, since the more channels being sampled, the larger the size of the transmitted data packets, and the higher

Table 4 Waveform period measurement and drift characterization results for the AX1 channel, using different ADC channel configurations

Number of active channels			Measured wave period statistics			Measured drift statistics	
External	Internal	Sampling rate (Hz)	Min (μs)	Max (μs)	Mean (μs)	Cumulative value (μs)	Compensation (μs/s)
1	0	10,000	9859	10,137	10,000	1597	7
	1	8000	9812	10,176	10,000	1750	7
	2	8000	9819	10,176	10,000	1753	7
	3	5000	9740	10,260	10,000	1607	7
	4	5000	9739	10,263	10,000	1557	6
	5	4000	9705	10,300	10,000	1466	6
	6	2000	9451	10,543	10,000	1507	6
2	0	5000	9761	10,231	10,000	1599	7
	1	5000	9748	10,250	10,000	1592	7
	2	4000	9708	10,291	10,000	1749	7
	3	2000	8996	10,531	10,000	1484	6
	4	2000	9457	10,542	10,000	1516	6
	5	2000	9481	10,557	10,000	1496	6
	6	2000	9456	10,544	10,000	1530	6

the sampling rate, the more data packets are sent per unit of time. As a result, there is a trade-off between the number of ADC channels sampled and the maximum achievable sampling rate.

Regarding the measure wave period statistics, the results show that, for both the internal and external ADC channels, the mean values coincide with the expected value for a 100 Hz square wave (10,000 μs), regardless of the sampling rate value or the channel configuration. This is also true for the median values, displayed in Tables 5 and 6.

Furthermore, the minimum and maximum values obtained do not deviate considerably from this expected value, creating a range of measured wave period values that is always centred around 10,000 μs. With this, the mean and median coinciding with the expected value and the fact that the standard deviation of the measured wave period is very low (in the order of magnitude of 10^{-5} s for most tested sampling rate values, as seen in Tables 5 and 6), it can be deduced that these values oscillate very closely to the expected value.

Regarding the measured drift statistics, the accuracy of the ADC process is once again reaffirmed by the measured cumulative value of the drift. These values vary very little across all tested conditions, for both the internal and external analog channels, ranging from around 1400 to 1800 μs, or 0.0014 to 0.0018 s, a very low value, especially when compared to the total duration of data acquisition (240 seconds). As mentioned before, these temporal inaccuracies in the ADC process are now solely inherent to the CORE board and no longer include those added by using the deduced timestamp instead of the real, which explains why there was a marked decrease in the cumulative value compared to the results obtained in CITE. While these temporal inaccuracies in the ADC process still accumulate, it is no cause for concern, as they do so very slowly, which limits their impact in the acquisition of biosignals.

Another reason why this drift is not detrimental to biosignal acquisition is the fact that its growth over time can be approximated to a linear function, meaning that timestamps can be regularized, if this function is known, to reduce the effect of drift during data acquisition. The slope of this function corresponds to the compensation per second metric, whose values are markedly smaller and less variable, across all tested conditions, than those obtained in CITE, since the drift measured here is only introduced by the CORE board itself, as was mentioned before.

It is important to note that Tables 3 & 4, as well as Tables 5, 6, 7 & 8 in Annex 1, comprise data from both the tests carried out using BT and Wi-Fi as the communication mode. This is because the statistics displayed in these tables evaluate the performance of the ADC process and no data packet loss was observed when using either communication mode, which means there should be no meaningful difference in the results obtained for either mode. To confirm this, we refer the reader to Tables 9 & 10 in Annex 1, which comprise the percentage change values of the measured wave period and drift statistics between the two modes, for an internal and external ADC channel, respectively.

The measured wave period results from these tables, particularly the mean and median, show practically no difference between BT and Wi-Fi, while there are higher percentage change values in the measured drift statistics for both internal and external channels, which was to be expected, taking into consideration these values' order of dimension (e.g. a minimum drift of $-1\,\mu$s compared to $-10\,\mu$s is a 900% decrease), but these have no real effect in the reliability of data acquisition.

On the other hand, for the wave period measurement and drift characterization results obtained when storing the data in an SD card, again no data packet loss was observed, which means that these results are once again a measure of the temporal accuracy of the ADC process. Thus, the results for the SD card are very similar to the results obtained when using BT or Wi-Fi, matching the expected values, regardless of the mode used (all internal analog channels sampled at 1 kHz, or all internal and external analog channels sampled at 100 Hz) or channel tested. To confirm this, we direct the reader to Tables 11 and 12.

A final note regarding the sampling rate values tested, it can be observed that these values are all factors of the base clock frequency of the ESP32 MCU module (i.e. 80 MHz). This is not by chance, as it was found in [24] that the accumulated drift in the ADC process could only be compensated properly, when the prescaler (p) meets the following condition: $p = \frac{80\,\mathrm{MHz}}{f_s} \in \mathbb{N}$; otherwise the timer for analog channel sampling, which works at this same base frequency, has to resort to approximations that introduce greater amounts of drift.

In sum, from the wave period measurements and drift characterization results obtained with different ADC channel configurations and sampling rates, it can be concluded that the CORE board is capable of accurately and reliably converting signals from the analog to the digital domain, irrespective of the communication mode used (BT, Wi-Fi, SD card). There was also a marked improvement in both drift measurements and maximum achievable sampling rate compared to the results in CITE, credited to the inclusion of the real timestamp of data acquisition in the transmitted data packets. However, there still needs to be some care in the selection of the sampling rate, to make sure that the ratio between the base clock frequency and the selected value is an integer, to make sure drift is kept to a minimum, and can be compensated in post-processing.

6 Summary

Current tools for physiological computing are lacking in many aspects that make them unsuitable for democratising rapid prototyping of applications which involve biosignal acquisition and processing. Serving as a new gold standard device for this field, ScientISST CORE was developed to support biosignal acquisition, improving accessibility and stimulating innovation. The following key points summarize the contents discussed throughout the chapter:

- **Physiological computing** has recently grown in popularity, in both academical and educational settings, but **existing low-cost hardware platforms** face **several issues** (e.g. discontinuation, limited functionality and applicability).
- **ScientISST CORE** is a reference low-cost hardware platform, designed to accelerate experimentation in biomedical engineering.
- This device's design includes **improved technical specifications** to overcome the existing limitations for biosignal acquisition in other devices.
- It can be **used in an out-of-the-box manner**, thanks to its custom-built **firmware optimized for real-time data acquisition and streaming**, as well as the ecosystem of tools built around it, which includes **sensors** and **programming APIs**.
- **Performance tests** were conducted to evaluate the **temporal accuracy of the ADC** process and **reliability of data transmission** using different communication modes (BT, Wi-Fi) and data storage using an SD card.
- Regardless of the communication mode used or if data was being stored in an SD, card, **no data packet loss was observed**, across all tested conditions.
- Results indicated that **real-time signal acquisition** was **accurate and reliable**, with statistics from the acquired signals coinciding with the expected values.
- **Signal drift also decreased compared to past performance tests**, due to the **inclusion of the real timestamp** in the transmitted data packets, allowing for **higher maximum achievable sampling rates** without data acquisition becoming unreliable.

Annex 1

Table 5 Waveform period measurement results for the AI1 channel, using different ADC channel configurations, for all sampling rate values tested

Number of active channels			Measured wave period statistics				
External	Internal	Sampling rate (Hz)	Min (μs)	Max (μs)	Mean (μs)	Median (μs)	Standard deviation (μs)
0	1	1000	9001	11,000	10,000	10,000	29
		2000	9500	10,500	10,000	10,000	25
		4000	9249	10,780	10,000	10,000	51
		5000	9798	10,202	10,000	10,000	19
		8000	9871	10,131	10,000	10,000	11
		10,000	9868	10,113	10,000	10,000	15
		16,000	9747	10,234	10,000	10,000	30
	2	1000	8998	11,002	10,000	10,000	26
		2000	9498	10,502	10,000	10,000	25
		4000	9733	10,268	10,000	10,000	22
		5000	9780	10,224	10,000	10.,000	24
		8000	7250	13,125	10,000	10,000	43
		10,000	9860	10,130	10,000	10,000	28
	3	1000	9006	11,000	10,000	10,000	51
		2000	9503	10,501	10,000	10,000	26
		4000	9745	10,255	10,000	10,000	28
		5000	9794	10,205	10,000	10,000	16
	4	1000	8996	11,007	10,000	10,000	41
		2000	9500	10,501	10,000	10,000	40
		4000	9711	10,285	10,000	10,000	20
		5000	9776	10,245	10,000	10,000	31
	5	1000	9002	11,003	10,000	10,000	51
		2000	9485	10,523	10,000	10,000	36
		4000	9729	10,260	10,000	10,000	31
	6	1000	9001	10,999	10,000	10,000	40
		2000	9499	10,501	10,000	10,000	38
		4000	9737	10,255	10,000	10,000	64
	1	1000	9017	10,984	10,000	10,000	35
		2000	9493	10,510	10,000	10,000	28
		4000	9741	10,259	10,000	10,000	23
		5000	9786	10,225	10,000	10,000	19
		8000	9869	10,147	10,000	10,000	19

(continued)

Table 5 (continued)

Number of active channels			Measured wave period statistics				
External	Internal	Sampling rate (Hz)	Min (μs)	Max (μs)	Mean (μs)	Median (μs)	Standard deviation (μs)
1	2	1000	8998	11,002	10,000	10,000	21
		2000	9488	10,512	10,000	10,000	42
		4000	9746	10,251	10,000	10,000	24
		5000	9800	10,202	10,000	10,000	25
		8000	9844	10,155	10,000	10,000	32
	3	1000	8999	11,000	10,000	10,000	28
		2000	9493	10,500	10,000	10,000	35
		4000	9749	10,253	10,000	10,000	24
		5000	9794	10,217	10,000	10,000	23
	4	1000	8987	11,014	10,000	10,000	55
		2000	9482	10,516	10,000	10,000	34
		4000	9743	10,266	10,000	10,000	29
		5000	9748	10,232	10,000	10,000	56
	5	1000	8996	11,003	10,000	10,000	31
		2000	9494	105,06	10,000	10,000	35
		4000	9720	10,283	10,000	10,000	43
	6	1000	8976	11,024	10,000	10,000	45
		2000	9494	10,512	10,000	10,000	46
2	1	1000	9000	10,999	10,000	10,000	21
		2000	9506	10,499	10,000	10,000	29
		4000	9748	10,269	10,000	10,000	28
		5000	9769	10,227	10,000	10,000	42
	2	1000	9016	10,999	10,000	10,000	16
		2000	9486	10,513	10,000	10,000	39
		4000	9713	10,286	10,000	10,000	47
	3	1000	8993	11,020	10,000	10,000	17
		2000	9490	10,507	10,000	10,000	30
	4	1000	8998	11,003	10,000	10,000	37
		2000	9484	10,517	10,000	10,000	29
	5	1000	8992	11,010	10,000	10,000	60
		2000	9497	10,501	10,000	10,000	40
	6	1000	8971	11,028	10,000	10,000	45
		2000	9491	10,513	10,000	10,000	44

Table 6 Waveform period measurement results for the AX1 channel, using different ADC channel configurations, for all sampling rate values tested

Number of active channels			Measured wave period statistics				
External	Internal	Sampling rate (Hz)	Min (μs)	Max (μs)	Mean (μs)	Median (μs)	Standard deviation (μs)
	0	1000	8995	11,498	10,000	10,000	156
		2000	9457	10,543	10,000	10,000	138
		4000	9708	10,292	10,000	10,000	91
		5000	9756	10,244	10,000	10,000	83
		8000	9844	10,156	10,000	10,000	67
		10,000	9859	10,137	10,000	10,000	57
	1	1000	8999	11,000	10,000	10,000	132
		2000	9484	10,517	10,000	10,000	122
		4000	9713	10,283	10,000	10,000	95
		5000	9755	10,237	10,000	10,000	84
		8000	9812	10,176	10,000	10,000	68
	2	1000	8993	11,008	10,000	10,000	137
		2000	9460	10,540	10,000	10,000	146
		4000	9711	10,289	10,000	10,000	92
1		5000	9785	10,219	10,000	10,000	85
		8000	9819	10,176	10,000	10,000	69
	3	1000	8971	11,029	10,000	10,000	133
		2000	9478	10,530	10,000	10,000	125
		4000	9704	10,293	10,000	10,000	89
		5000	9740	10,260	10,000	10,000	84
	4	1000	8973	11,021	10,000	10,000	189
		2000	9468	10,531	10,000	10,000	130
		4000	9704	10,293	10,000	10,000	95
		5000	9739	10,263	10,000	10,000	87
	5	1000	8992	11,007	10,000	10,000	142
		2000	9475	10,525	10,000	10,000	122
		4000	9705	10,300	10,000	10,000	91
	6	1000	8976	11,025	10,000	10,000	146
		2000	9451	10,543	10,000	10,000	128
	0	1000	7998	12,002	10,000	10,000	261
		2000	9468	10,532	10,000	10,000	149
		4000	9500	10,749	10,000	10,000	106
		5000	9761	10,231	10,000	10,000	97
	1	1000	8970	11,015	10,000	10,000	152
		2000	8519	11,482	10,000	10,000	144
		4000	9714	10,282	10,000	10,000	103
		5000	9748	10,250	10,000	10,000	98

(continued)

Table 6 (continued)

Number of active channels			Measured wave period statistics				
External	Internal	Sampling rate (Hz)	Min (μs)	Max (μs)	Mean (μs)	Median (μs)	Standard deviation (μs)
2	2	1000	8991	11,010	10,000	10,000	154
		2000	9472	10,528	10,000	10,000	145
		4000	9708	10,291	10,000	10,000	111
	3	1000	8958	11,041	10,000	10,000	192
		2000	8996	10,531	10,000	10,000	144
	4	1000	8984	11,011	10,000	10,000	221
		2000	9457	10,542	10,000	10,000	146
	5	1000	8946	11,042	10,000	10,000	218
		2000	9481	10,557	10,000	10,000	141
	6	1000	8962	11,038	10,000	10,000	232
		2000	9456	10,544	10,000	10,000	148

Table 7 Drift characterization results for the AI1 channel, using different ADC channel configurations, for all sampling rate values tested

Number of active channels			Measured drift statistics					
External	Internal	Sampling rate (Hz)	Min (μs)	Max (μs)	Mean (μs)	Median (μs)	Standard deviation (μs)	Cumulative value (μs)
0	1	1000	−7	2035	1053	994	578	1994
		2000	−4	1540	783	1000	494	1500
		4000	−172	1889	594	580	391	1146
		5000	0	1625	816	800	476	1601
		8000	−3	1544	742	750	444	1505
		10,000	−18	1626	821	801	473	1600
		16,000	−49	1663	803	794	475	1620
	2	1000	0	2036	828	1000	565	2002
		2000	−1	1530	793	1000	495	1500
		4000	−249	1534	691	749	471	1500
		5000	0	1614	873	808	467	1601
		8000	−102	3492	610	617	395	1368
		10,000	−82	1639	781	783	476	1581
	3	1000	−9	2003	920	994	586	1994
		2000	−6	1557	751	996	497	1524
		4000	−1	1530	729	752	474	1505
		5000	0	1811	887	805	471	1607
	4	1000	−13	2028	1004	998	583	1990
		2000	0	2009	955	1000	2000	7
		4000	−6	1539	769	757	474	1496
		5000	−17	1633	791	809	479	1603
	5	1000	−1	1059	612	1000	488	1000
		2000	0	1551	676	507	496	1517
		4000	−2	1793	924	998	477	1756
	6	1000	−935	1052	366	6	482	1008
		2000	−8	1539	621	498	482	1492
		4000	−30	1846	909	894	478	1717
	1	1000	−10	1031	575	991	488	991
		2000	−3	1517	590	503	464	1500
		4000	−3	1811	886	1000	472	1750
		5000	−1	1634	772	801	478	1605
		8000	−13	1663	833	875	478	1625
	2	1000	−2	1049	499	19	501	1000
		2000	−5	2044	998	995	468	2007
		4000	−230	1537	697	750	472	1500
		5000	−157	1641	732	800	476	1600
		8000	−10	1795	870	869	484	1753

(continued)

Table 7 (continued)

Number of active channels			Measured drift statistics					
External	Internal	Sampling rate (Hz)	Min (μs)	Max (μs)	Mean (μs)	Median (μs)	Standard deviation (μs)	Cumulative value (μs)
1	3	1000	−3	1032	376	5	483	1031
		2000	−8	1536	862	993	485	1493
		4000	−202	1560	710	750	473	1500
		5000	−153	1641	756	800	479	1607
	4	1000	−3	2046	1172	998	579	2010
		2000	−2	1535	868	1001	487	1500
		4000	−11	1549	775	752	478	1503
		5000	−24	1808	898	939	482	1752
	5	1000	−3	1045	480	10	501	1001
		2000	−59	1493	729	947	497	1451
		4000	−43	1518	706	719	476	1466
	6	1000	−13	1033	367	9	477	993
		2000	−5	1550	739	506	505	1507
2	1	1000	−4	1034	490	6	500	998
		2000	−23	1513	878	977	482	1477
		4000	−1	1542	759	751	476	1501
		5000	−23	1815	886	826	474	1788
	2	1000	−4	1042	457	8	497	1009
		2000	−2	1529	868	1002	486	1502
		4000	−19	1792	878	982	483	1749
	3	1000	−2	1069	393	0	495	1016
		2000	−23	1517	715	489	504	1484
	4	1000	−2	2045	1224	1003	565	2001
		2000	−3	1567	816	998	506	1516
	5	1000	−25	2022	972	994	597	1979
		2000	−5	1529	754	996	502	1496
	6	1000	−1	2050	818	1006	576	2005
		2000	−3	1558	653	507	496	1530

Table 8 Drift characterization results for the AX1 channel, using different ADC channel configurations, for all sampling rate values tested

Number of active channels			Measured drift statistics					
External	Internal	Sampling rate (Hz)	Min (μs)	Max (μs)	Mean (μs)	Median (μs)	Standard deviation (μs)	Cumulative value (μs)
1	0	1000	−428	2739	701	1000	521	2000
		2000	−5	2030	1048	996	465	1996
		4000	−1	1529	777	750	475	1500
		5000	−3	1612	810	800	478	1600
		8000	−2	1654	802	758	472	1625
		10,000	−103	1698	793	797	474	1597
	1	1000	−10	1031	596	991	485	991
		2000	−3	1517	621	504	471	1500
		4000	−258	1553	664	742	469	1492
		5000	−1	1653	804	803	478	1605
		8000	−13	1786	868	875	478	1750
	2	1000	−2	1049	521	1000	501	1000
		2000	−500	1549	548	500	469	1512
		4000	−1	1537	728	750	475	1500
		5000	−1	1641	767	800	479	1600
		8000	−8	1795	904	876	484	1753
	3	1000	−3	1032	399	6	489	1031
		2000	−8	1536	894	993	477	1493
		4000	−1	1560	742	750	475	1500
		5000	−13	1642	790	800	481	1607
	4	1000	−3	2046	1214	998	568	2010
		2000	−2	1535	902	1002	478	1500
		4000	−11	1773	809	756	478	1503
		5000	−213	1629	738	747	485	1557
	5	1000	−3	1045	502	22	502	1001
		2000	−59	1493	758	952	494	1451
		4000	−43	1723	735	722	477	1466
	6	1000	−13	1034	389	10	483	993
		2000	−5	1550	772	997	506	1507
	0	1000	−462	3000	531	1000	507	1000
		2000	−6	2004	907	996	482	1496
		4000	−250	1752	741	751	478	1500
		5000	−3	1829	868	801	481	1599
	1	1000	−4	1034	480	3	500	998
		2000	−23	1981	864	977	487	1477
		4000	−250	1759	744	750	478	1501
		5000	−23	1815	870	813	481	1592

(continued)

Table 8 (continued)

Number of active channels			Measured drift statistics					
External	Internal	Sampling rate (Hz)	Min (μs)	Max (μs)	Mean (μs)	Median (μs)	Standard deviation (μs)	Cumulative value (μs)
2	2	1000	−4	1042	447	7	496	1009
		2000	−2	1529	852	1002	489	1502
		4000	−19	1787	861	761	482	1749
	3	1000	−29	2041	1342	972	510	1988
		2000	−23	1517	700	486	502	1484
	4	1000	−2	2045	1203	1003	572	2001
		2000	−3	1567	801	998	508	1516
	5	1000	−25	2022	949	994	595	1979
		2000	−5	1529	739	525	502	1496
	6	1000	−1	2050	798	1006	569	2005
		2000	−500	1558	638	506	493	1530

Table 9 Comparison between waveform period measurement and drift characterization results for the AI1 channel, using different ADC channel configurations, when using BT or Wi-Fi for data transfer (in %)

Number of active channels		Sampling rate (Hz)	Measured wave period statistics				Measured drift statistics				
External	Internal		Min(μs)	Max (μs)	Mean (μs)	Median (μs)	Min (μs)	Max (μs)	Mean (μs)	Median (μs)	Cumulative value (μs)
0	1	1000	0.64	−0.01	0.00	0.00	100.00	−33.51	1.04	−0.80	−0.35
		2000	5.31	−2.38	0.00	0.00	−1175.00	−50.52	−16.73	9.90	−28.07
		10,000	−0.28	0.08	0.00	0.00	94.44	1.05	3.29	0.12	−0.25
		16,000	−1.68	1.06	0.00	0.03	−16.33	5.65	6.10	3.78	5.80
	2	1000	−0.02	0.02	0.00	0.00	inf	−29.13	−57.61	−0.20	0.10
		2000	5.29	−2.74	0.00	0.00	−14300.00	−48.17	2.40	22.70	−13.47
		10,000	−0.17	0.07	0.00	0.00	91.46	0.73	−4.23	−2.30	−2.34
	3	1000	0.84	−0.05	0.00	0.00	77.78	14.43	36.63	−0.60	49.70
		2000	5.30	−1.68	0.00	0.00	−66.67	−54.66	−31.56	−2.61	−29.40
		4000	−0.01	0.01	0.00	0.00	100.00	−14.51	−17.01	−0.27	−16.28
	4	1000	−0.04	−5.24	0.00	0.00	92.31	−33.97	−12.95	−0.30	−0.75
		2000	5.28	−2.28	0.00	0.00	inf	−1.74	26.39	27.50	29.40
		5000	−0.09	0.29	0.00	0.00	−5.88	0.49	−3.79	1.11	0.19
	5	1000	0.02	0.03	0.00	0.00	−68200.00	−64.97	27.94	100.10	0.10
		2000	−0.14	−2.12	0.00	0.00	inf	−33.40	8.73	21.50	12.00
		4000	−0.16	0.05	0.00	0.00	−400.00	15.56	29.65	24.95	14.64
	6	1000	0.06	0.00	0.00	0.00	99.89	−49.52	−74.32	−16566.67	0.69
		2000	5.32	−2.16	0.00	0.00	−1312.50	−43.27	−40.58	−84.54	−22.65

(continued)

Table 9 (continued)

Number of active channels		Sampling rate (Hz)	Measured wave period statistics				Measured drift statistics				
External	Internal		Min(µs)	Max (µs)	Mean (µs)	Median (µs)	Min (µs)	Max (µs)	Mean (µs)	Median (µs)	Cumulative value (µs)
1	1	1000	0.19	−0.15	0.00	0.00	−4300.00	−69.06	−4.87	−0.91	−1.01
		4000	−0.09	0.09	0.00	0.00	66.67	15.35	13.54	25.00	14.29
	2	1000	0.19	0.03	0.00	0.00	−26200.00	−67.21	0.00	−173.68	−0.30
		2000	5.13	−2.32	0.00	0.00	−8100.00	15.02	45.89	36.78	40.91
		5000	0.08	−0.12	0.00	0.00	98.09	13.65	6.69	24.62	12.56
		8000	−0.07	0.03	0.00	0.00	80.00	14.32	10.69	12.54	14.26
	3	1000	−0.01	0.00	0.00	0.00	−29866.67	−64.05	5.32	140.00	3.49
		4000	0.00	0.01	0.00	0.00	100.00	−12.95	−24.93	−33.33	−16.67
	4	1000	−0.16	−5901.41	−0.59	0.00	33.33	−34242.23	−13104.10	−0.20	−34825.32
		2000	5.11	−1.94	0.00	0.00	−1400.00	−55.57	−4.61	9.49	−26.00
		5000	−0.47	0.26	0.00	0.00	83.33	6.97	6.79	14.38	8.62
	5	1000	−0.11	0.04	0.00	0.00	−26000.00	−62.68	3.96	90.00	0.70
	6	1000	−0.22	0.21	0.00	0.00	84.62	−153.15	−237.60	−11011.11	−102.01
		2000	−0.08	0.05	0.00	0.00	60.00	2.32	15.02	1.19	0.53
2	1	1000	0.08	−1.49	0.00	0.00	−23800.00	−65.86	24.08	83.33	−0.20
		2000	5.39	−2.08	0.00	0.00	−52.17	−54.66	−9.45	−0.41	−6.84
	2	1000	0.18	−0.03	0.00	0.00	75.00	−156.91	−176.59	−12425.00	−98.51
		4000	−0.12	0.13	0.00	0.00	5.26	13.50	7.40	23.63	15.09
	3	1000	−0.01	0.13	0.00	0.00	50.00	−154.16	−171.76	-inf	−96.85
		2000	5.22	−1.93	0.00	0.00	−1708.70	−44.36	16.50	−41.51	25.81
	4	1000	−0.02	0.02	0.00	0.00	−31350.00	17.90	69.20	99.50	50.02
	5	1000	−0.03	0.05	0.00	0.00	100.00	−14.69	−5.97	−0.60	−1.06
		2000	−0.03	−0.02	0.00	0.00	60.00	0.98	28.12	49.80	−0.20
	6	1000	−0.32	0.24	0.00	0.00	−700.00	30.88	48.53	99.80	50.17

Table 10 Comparison between waveform period measurement and drift characterization results for the AX1 channel, using different ADC channel configurations, when using BT or Wi-Fi for data transfer (in %)

Number of active channels		Sampling rate (Hz)	Measured wave period statistics				Measured drift statistics				
External	Internal		Min (μs)	Max (μs)	Mean (μs)	Median (μs)	Min (μs)	Max (μs)	Mean (μs)	Median (μs)	Cumulative value (μs)
1	0	2000	4.77	−1.93	0.00	0.00	−9900.00	1.63	49.33	47.59	38.28
		4000	−0.40	0.37	0.00	0.00	−100.00	−6.28	0.13	−16.13	−8.13
		5000	−0.39	0.37	0.00	0.00	66.67	0.31	6.17	0.00	0.00
		8000	−0.23	0.19	0.00	0.00	50.00	1.45	1.37	0.92	0.00
		10,000	−0.32	0.27	0.00	0.00	1.94	11.19	7.82	12.17	6.07
	1	1000	0.06	−0.07	0.00	0.00	−4300.00	−69.06	−5.37	−0.91	−1.01
		4000	−0.29	0.23	0.00	0.00	99.61	1.29	−20.93	−1.08	−0.54
	2	1000	0.13	−0.03	0.00	0.00	−26200.00	−67.21	−0.77	0.00	−0.30
		2000	4.85	−2.05	0.00	0.00	18.00	−12.14	−5.66	−62.20	21.56
		5000	−0.01	0.08	0.00	0.00	−200.00	2.25	5.74	0.12	12.56
		8000	−0.26	0.25	0.00	0.00	−1750.00	15.88	25.88	27.85	15.35
	3	1000	−0.31	0.24	0.00	0.00	−29866.67	−64.05	4.51	116.67	3.49
		4000	−0.34	0.28	0.00	0.00	−25000.00	3.08	8.89	0.13	0.07
	4	1000	−0.26	−5888.52	−0.59	0.00	33.33	−34242.23	−12651.40	−0.20	−34825.32
		2000	5.02	−1.79	0.00	0.00	−1400.00	−55.57	−6.21	5.39	−26.00
		5000	−0.49	0.20	0.00	0.00	98.12	−3.25	−18.70	−12.32	−2.83
	5	1000	−0.06	−0.04	0.00	0.00	−23500.00	−62.68	3.19	86.36	0.70
	6	1000	−0.13	0.13	0.00	0.00	84.62	−156.00	−231.88	−9900.00	−102.01
		2000	−0.43	0.32	0.00	0.00	60.00	2.26	13.60	49.85	0.53

(continued)

Table 10 (continued)

Number of active channels		Sampling rate (Hz)	Measured wave period statistics				Measured drift statistics				
External	Internal		Min (μs)	Max (μs)	Mean (μs)	Median (μs)	Min (μs)	Max (μs)	Mean (μs)	Median (μs)	Cumulative value (μs)
2	0	4000	0.03	4.54	0.00	0.00	99.60	−0.11	−16.73	−0.13	−16.87
		5000	1.68	−1.70	0.00	0.00	33.33	1.59	0.58	0.25	0.00
	1	1000	10.82	−8.95	0.00	0.00	−24925.00	−93.52	24.79	66.67	−0.20
		2000	−5.54	3.43	0.00	0.00	−417.39	−21.86	−10.65	1.13	−6.84
	2	1000	11.06	−9.04	0.00	0.00	75.00	−188.39	−180.09	−14214.29	−98.51
		4000	2.31	−2.03	0.00	0.00	5.26	1.79	6.85	1.45	−0.06
	3	1000	10.76	−8.75	0.00	0.00	96.55	−47.43	21.09	−2.88	−0.60
		2000	0.01	−4.47	0.00	0.00	−2073.91	−44.36	16.43	−35.39	25.81
	4	1000	11.01	−8.99	0.00	0.00	−49900.00	1.91	69.33	99.50	50.02
	5	1000	10.59	−8.69	0.00	0.00	100.00	−48.76	−7.38	−0.60	−1.06
		2000	10.35	−8.92	0.00	0.00	80.00	−96.14	−39.11	−90.48	−33.62
	6	1000	10.80	−8.77	0.00	0.00	−700.00	−4.49	47.49	99.90	50.17

Table 11 Waveform period measurement results for the AI1 and AX1 channel, using different ADC channel configurations, in SD card mode

Number of Active channels				Measured wave period statistics				
External	Internal	Channel	Sampling rate (Hz)	Min (μs)	Max (μs)	Mean (μs)	Median (μs)	Standard deviation (μs)
0	6	AI1	1000	9984	10,016	10,000	10,000	3
2	6	AI1	100	91,168	108,833	100,000	100,000	530
2	6	AX1	100	91,082	108,918	100,000	100,000	590

Table 12 Drift characterization results for the AI1 and AX1 channel, using different ADC channel configurations, in SD card mode

Number of active channels				Measured drift statistics					
External	Internal	Channel	Sampling rate (Hz)	Min (μs)	Max (μs)	Mean (μs)	Median (μs)	Standard deviation (μs)	Cumulative value (μs)
0	6	AI1	1000	−2	16	1	0	2	0
2	6	AI1	100	−8833	6526	−11	−1	375	0
2	6	AX1	100	−8919	6526	−14	−1	417	0

Acknowledgments This work was partially funded by FCT/MECI through national funds and when applicable co-funded EU funds under project UID/50008: Instituto de Telecomunicações. It was also co-supported by the European Regional Development Fund (FEDER), through the Lisbon Regional Programme (LISBOA 2030) of the Portugal 2030 framework—LISBOA2030-FEDER-01318700 (ComSense).

References

1. D. Cressey, The DIY electronics transforming research. Nature **544**, 125–126 (2017)
2. S.S. Barold, Willem Einthoven and the birth of clinical electrocardiography a hundred years ago. Cardiac Electrophysiol. Rev. **7**(1), 99–104 (2003)
3. A. Sharma, R.A. Harrington, M.B. McClellan, M.P. Turakhia, Z.J. Eapen, S. Steinhubl, J.R. Mault, M.D. Majmudar, L. Roessig, K.J. Chandross, et al., Using digital health technology to better generate evidence and deliver evidence-based care. J. Am. College Cardiol. **71**, 2680–2690 (2018)
4. S.P. Bhavnani, J. Narula, P.P. Sengupta, Mobile technology and the digitization of Healthcare. Eur. Heart J. **37**, 1428–1438 (2016)
5. H. Turker, H. Sze, Surface electromyography in sports and exercise. Electrodiagn. New Front. Clin. Res. **2013**, 175–194 (2013)
6. I.M. Mercado-Aguirre, E.L. Mercado-Medina, Z.D. Chavarro-Hernandez, J.A. Dominguez-Jimenez, S.H. Contreras-Ortiz, A wearable system for biosignal monitoring in weightlifting. Sports Eng. **20**, 73–80 (2016)
7. N.N. Correia, R. Masu, W. Primett, S. Jürgens, J. Feitsch, H. Plácido da Silva, Designing interactive visuals for dance from body maps: machine learning and composite animation approaches, in *Proceedings of the 2022 ACM Designing Interactive Systems Conference*, DIS '22, (New York, NY, USA) (Association for Computing Machinery, 2022), pp. 204–216

8. A. Kaur, Wheelchair control for disabled patients using EMG/EOG based human machine interface: a Review. J. Med. Eng. Technol. **45**, 61–74 (2020)
9. H.P. da Silva, A. Fred, A. Lourenço, A.K. Jain, Finger ECG signal for user authentication: Usability and performance, in *2013 IEEE Sixth International Conference on Biometrics: Theory, Applications and Systems (BTAS)* (2013), pp. 1–8
10. A. Lourenço, H. Silva, A. Fred, Unveiling the biometric potential of finger-based ECG signals. Comput. Intell. Neurosci. **2011**, 1–8 (2011)
11. C. Krittanawong, A.J. Rogers, K.W. Johnson, Z. Wang, M.P. Turakhia, J.L. Halperin, S.M. Narayan, Integration of novel monitoring devices with machine learning technology for scalable cardiovascular management. Nat. Rev. Cardiol. **18**, 75–91 (2020)
12. G. Ehrmann, T. Blachowicz, S.V. Homburg, A. Ehrmann, Measuring biosignals with single circuit boards. Bioengineering **9**, 84 (2022)
13. M. Banzi, M. Shiloh, *Getting started with Arduino* (Make Community, 2022)
14. P. Bender, K. Kussmann, Arduino based projects in the computer science capstone course. J. Comput. Sci. Colleges **27**, 152–157 (2012)
15. J. Sarik, I. Kymissis, Lab kits using the Arduino Prototyping platform, in *2010 IEEE Frontiers in Education Conference (FIE)* (2010)
16. J.F. Pinto, H.P. Silva, F. Melo, A. Fred, ScientIST: biomedical engineering experiments supported by mobile devices, cloud and IoT. Signals **1**, 110–120 (2020)
17. J. Guerreiro, A. Lourenço, H. Silva, A. Fred, Performance comparison of low-cost hardware platforms targeting physiological computing applications. Proc. Technol. **17**, 399–406 (2014)
18. OpenEEG Project (2018). https://openeeg.sourceforge.net/doc/hw/
19. V. Peterson, C. Galván, H. Hernández, R. Spies, A feasibility study of a complete low-cost consumer-grade brain-computer interface system. Heliyon **6**, e03425 (2020)
20. H. Plácido da Silva, J. Guerreiro, A. Lourenço, A. Fred, R. Martins, BITalino: a novel hardware framework for physiological computing, in *Proceedings of the International Conference on Physiological Computing Systems*, (Lisbon, Portugal), SCITEPRESS—Science and and Technology Publications (2014), pp. 246–253
21. D. Batista, H. Silva, A. Fred, Experimental characterization and analysis of the BITalino platforms against a reference device, in *2017 39th Annual International Conference of the IEEE Engineering in Medicine and Biology Society (EMBC)* (2017), pp. 2418–2421
22. H. Naseri, M.R. Homaeinezhad, Detection and boundary identification of phonocardiogram sounds using an expert frequency-energy based metric. Ann. Biomed. Eng. **41**, 279–292 (2012)
23. H. Zumbahlen, Staying well grounded (2012). Accessed 9 Apr 2022
24. L. Pereira, F. Melo, F.A. Santos, A. Ferreira, H.P. da Silva, ScientISST CORE: a novel hardware development platform for biomedical engineering. HardwareX **21**, e00630 (2025). https://doi.org/10.1016/j.ohx.2025.e00630

Abstracting Low-Level Hardware and Firmware Specificities Through Application Programming Interfaces (APIs)

Frederico de Almeida Santos and Hugo Plácido da Silva

1 Introduction

Embedded systems, such as the ScientISST SENSE CORE, hereinafter referred to as CORE, typically operate using custom-developed firmware for data acquisition. This is motivated by the fact that, when transmission of the acquired data is needed, there is a wide range of high-level protocols that can be used [1, 2]; however, they can introduce significant temporal and spatial overheads prone to hindering the performance [3]. This can be particularly problematic for time-critical applications, motivating the development of more low-level solutions [4].

On the other hand, interfacing end-user applications directly with custom firmware can be challenging, due to the minutia and specialized knowledge required. Consequently, this chapter presents the principles underlying the creation of Application Programming Interfaces (APIs), using the ScientISST SENSE Python API as a case study. An API provides an intermediate layer between the firmware and the application-level software, enabling users to interact with the hardware devices through API function calls that encapsulate the intricacies of the firmware running on the devices. This abstraction not only benefits those unfamiliar with embedded systems, but also facilitates performance optimizations and future improvements to the communication layer without disrupting existing applications [5].

The programming language selected for this API was Python [6]. As an open-source, interpreted, high-level, and general-purpose language, Python supports procedural, object-oriented, and functional programming paradigms. One of its core design principles is code readability, facilitated by its dynamic typing and automatic memory management features. Additionally, Python is widely compatible

F. de Almeida Santos (✉) · H. P. da Silva
Instituto Superior Técnico, Universidade de Lisboa, Lisbon, Portugal

IT - Instituto de Telecomunicações, Lisbon, Portugal

H. P. da Silva et al. (eds.), *Open Source Biomedical Engineering*,
https://doi.org/10.1007/978-3-032-03655-1_3

with various chip architectures and operating systems. For the field of biomedical engineering, Python is particularly valuable due to the availability of open-source libraries like BioSPPy, `SciPy` [7], or `scikit-learn` [8], which enable low-level functions such as numerical data handling and processing, forming the foundation for physiological computing applications.

The following sections present the ScientISST SENSE Python API from a development standpoint. Firstly, the role and goals of the API are clearly defined, because those choices will shape the design. In this case, ease of use and abstraction from the embedded systems domain were prioritized, while ensuring that performance is not sacrificed, given the real-time requirements of physiological acquisition. Subsequently, some crucial implementation details are analyzed to showcase both how to interface with this class of devices and how to abstract that complexity for the API users. Finally, a sample application `sense.py` is also described to demonstrate the API's capabilities; this command-line script facilitates data acquisition from a CORE hardware device, generating a human-readable `.csv` file for further analysis and providing a real-time plot for visualizing the acquired data.

2 API Design

2.1 Overview

This section aims to explain the design of the API, starting with the general functionality, then specifying all the tasks the API is responsible for, and, finally, discussing the more critical choices regarding the implementation details. Some code snippets have been slightly changed to better fit as an example without changing the functionality.

Figure 1 shows an interaction diagram between a CORE hardware device, the API, and an application (`sense.py`) using Bluetooth communication, corresponding to the following code snippet.

```
BLUETOOTH_ADDRESS = "08:3A:F2:49:AB:DE"
scientisst = ScientISST(BLUETOOTH_ADDRESS) # Create ScientISST
    ↪ instance and connect to device

version = scientisst.version_and_adc_chars() # Obtain the device
    ↪ characteristics and return the firmware version

scientisst.start(100, [1]) # Start acquisition with 100hz sampling
    ↪ rate and only channel AI1

frames = [] # List of batches (list) of frames
# Read batches of frames in loop for the desired amount of time
for i in range(5):
    temp =  scientisst.read() # Read a batch of unpacked frames
```

```
12         frames.append(temp) # Append to list of batch of frames
13
14     scientisst.stop() # Stop acquisition
15     scientisst.disconnect() # Disconnect from device
```

The role of the API is clear when looking at both the code snippet and the diagram; it abstracts all the implementation details of the firmware. The user only has to call for the needed functionality—connect to the device, start acquisition (with given parameters), get unpacked frames, stop acquisition, and disconnect from the device—in an intuitive way and with a minimal amount of lines of code written. To build an application with the API, there is no need to know how the device's firmware works, how to use Bluetooth to communicate, how to unpack the frames sent by the device, or how to validate that the received frames have arrived correctly.

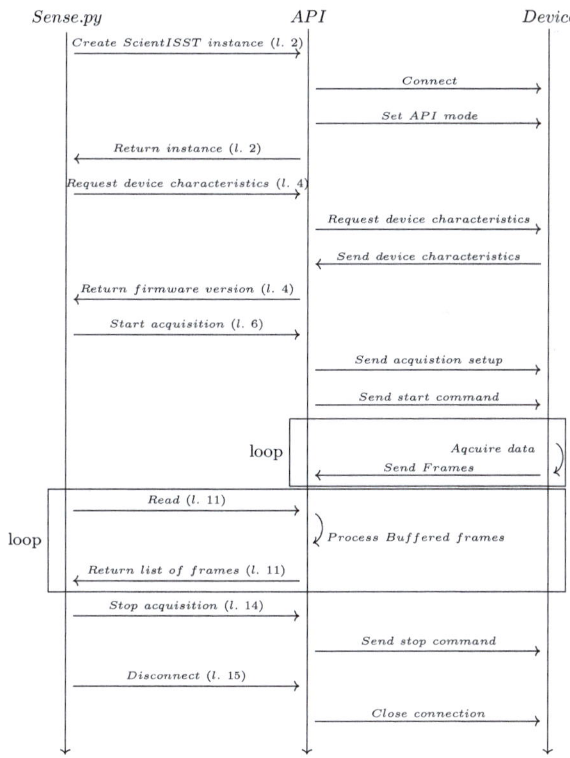

Fig. 1 Interactions diagram between an application (in this case `sense.py`), the API, and a ScientISST CORE device. The indicated line numbers (L) correspond to code lines of the previous code snippet

2.2 API Functionality

Having provided clarity on the role of the API, the next step is understanding all the tasks the API has to do to ensure it can fulfill its role. The API implements the following functionalities:

1. **Connection with device** (`Scientisst()`): Open a socket for the correct communication mode and connect with the device.
2. **Set API mode** (`Scientisst()`): The API is designed to work with all versions of the hardware and firmware, while ensuring older applications continue to function. This is achieved through different API modes, where each mode represents a different version. By default, all devices use the oldest mode to ensure maximum compatibility. If users want to access newer features, they can request the API to switch to a newer mode, although this will only work if the connected device supports these newer features. This approach ensures that updating the hardware or firmware will not break previously existing applications, while still allowing users to take advantage of new capabilities when available.
3. **Get device information** (`version_and_adc_chars()`): Three sets of values are requested from the devices—firmware version, Vref, and the analog-to-digital converter (ADC) characteristics. The firmware version is useful for debugging and to know which features are supported on the device. Vref is the current reference voltage used by the ADC. It is used in conjunction with the measured ADC characteristics to convert the raw data from the ADCs to mV.
4. **Send "acquisition setup" and start an acquisition** (`start()`): The API has to send the desired acquisition setup (which channels to acquire data from and at which sampling rate) and then send a "start" command to begin real-time data acquisition and streaming.
5. **Receive, unpack, and process frames** (`read()`): The operating system will receive data on the communication socket in the background and put it in a buffer. The API needs a function to read from that buffer, unpack the frames, and return them to the application. To save both bandwidth and processing time on the limited resources of CORE hardware devices, the data is sent in bit-packed frames following a predetermined protocol. The API has to unpack all frames, check for consistency by computing the Cyclic Redundancy Check (CRC) bits of all frames, and convert the raw data to mV so that is can be interpreted.
6. **Stop acquisition** (`stop()`): A command is needed to stop the acquisition and put the device in idle mode, waiting for the next command.
7. **Disconnect** (`disconnect()`): Disconnect gracefully from the device by sending the "disconnect" command to the device that will then pass to the "awaiting connection" mode and close the socket.
8. **Set digital output channels** (`trigger()`): The CORE devices have 2 digital channels that can be individually set to ON or OFF at any point. This

functionality is available at all times, regardless of being in real-time data acquisition or standby modes.

9. **Set digital-to-analog converter (DAC) level** (`dac()`): The CORE hardware devices have a DAC, and this command sets it to the requested output.
10. **Set battery threshold** (`battery()`): Set the threshold value that will be used to decide if the device has a low battery and turn on the corresponding low-battery LED.

2.3 Efficient Communication Between the API and the Devices

Figure 2 showcases the structure of the commands sent by the API to CORE devices. The commands occupy only between 1 and 3 bytes, which is important when dealing with limited-resource devices like the CORE. This is achieved by bit-packing [9] commands in a similar way to that used to pack the acquired data into frames. The concept of bit-packing is a fundamental aspect, and this section is dedicated to explain its importance and implementation.

Bit packing is a data compression method that reduces the number of bits required to represent a value to the bare minimum by eliminating unnecessary bits. This is a valuable technique when the data can have a very strict structure, like the

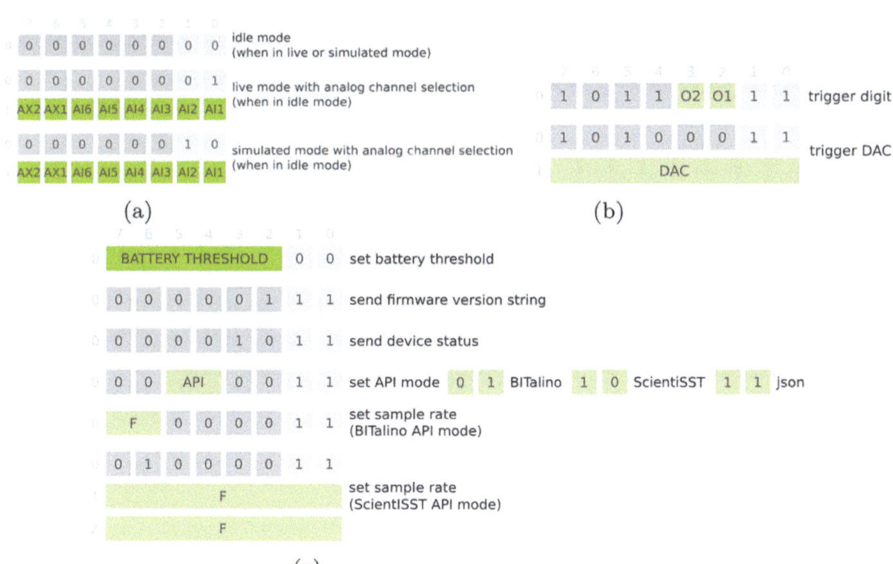

Fig. 2 Bit-level structure of the commands implemented in the CORE hardware device firmware and that the API encapsulates. (**a**) Operation commands. (**b**) Trigger commands. (**c**) Configuration commands

Fig. 3 Bit-packed structure of data frames. (**a**) One internal channel. (**b**) All internal channels. (**c**) All internal and external channels. (**d**) All internal and external channels using ScientisstV2 mode, which uses a timestamp instead of a sequence number

commands and the frames sent by CORE hardware devices (e.g., the internal ADC is a 12-bit number, the CRC is 4 bits, the digital channels are 1 bit—i.e., ON or OFF). For example, ADC values may have a precision of 12 bits, but using a 32-bit integer to store them and transmitting such data directly would waste 20 bits. Using bit-packing methods, it is possible to have a more efficient way of storing and sending data to and from the device to the API. This means that the device can store locally more frames before running out of memory and maximizes the transmission rate. Available memory is a restrictive constraint for this class of devices, making this step absolutely crucial.

For example, an acquisition with only one internal channel creates a frame of 4 bytes (Fig. 3) when bit-packed. On the other hand, if every parameter used 32-bit integers, the same frame would require 28 bytes (a 7× increase).

The process of unpacking is identical to the bit-packing process but reversed. The code snippet below is an example of how the API can unpack the sequence number from a frame sent by the device. It is important to note that the API knows which channels are active, since it sent that configuration to the device.

```
# Assume 16-bit integers for lower, upper, and seq_num
# packed_frame is an array of bytes
# Set original sequence number as: 0 0 0 0 s11 s10 s9 s8 s7 s6 s5 s4
↪   s3 s2 s1 s0
# It would get bit-packed inside a frame sent by the device
# The sequence number occupies the last byte and the 4 most
↪   significant bits (4,5,6,7) of the second-to-last byte of the
↪   frame
```

```python
# Last 2 bytes are organized as:
# Byte | Bits
# (-2) | s3 s2 s1 s0 c3 c2 c1 c0
# (-1) | s11 s10 s9 s8 c7 c6 c5 c4
lower = packed_frame[-2] >> 4 # lower = 0 0 0 0 0 0 0 0 0 0 0 0 s3 s2 s1 s0
upper = packed_frame[-1] << 4 # upper = 0 0 0 0 s11 s10 s9 s8 s7 s6 s5 s4 0 0 0 0
seq_num = upper | lower # seq_num = 0 0 0 0 s11 s10 s9 s8 s7 s6 s5 s4 s3 s2 s1 s0
```

2.4 *Ensuring the Validity of the Frames with CRC*

Every system must be resilient to errors, and the communication between a device and an API is no exception. In order to detect certain errors (e.g., the frames do not have the expected format, losses during transmission, API mode mismatch, acquisition setup mismatch, faulty firmware or device), we chose to include in every frame a 4-bit CRC (i.e., CRC-4) [10]; this enables the confirmation that a given unpacked frame is valid and identical to the frame sent by the device. CRC-4 is a method used to detect errors in transmitted data by calculating checksum values on 4-bit blocks of the data being sent. This checksum is then appended to the data and transmitted along with it. Finally, when the data is received, the same CRC calculation is performed on the received data. If the calculated checksum matches the received checksum, the data is assumed to be error-free; otherwise, an error is flagged. The downside of this method is that, while being lightweight in processing and memory, it does not provide any information on where the error is or its origin. Furthermore, validating that a data frame is received correctly does not mean the data values themselves are accurate. The actual data values need to be verified through additional analysis after collection.

The code snippet below showcases how the CRC validation process can be implemented. The creation and validation procedures of a CRC are identical, with the only difference being that one appends that value to the frame, and the latter checks that the value calculated is identical to the one on the frame.

```python
# The CRC table values
CRC4tab = [0, 3, 6, 5, 12, 15, 10, 9, 11, 8, 13, 14, 7, 4, 1, 2]
crc = 0
# Compute CRC for each byte except the last 2 because the second-to-last byte contains the CRC value, and it should not be included in the calculations
for i in range(length - 2):
    b = packed_frame[i] # Get byte i
    crc = CRC4tab[crc] ^ (b >> 4) # Compute CRC for 4 most significant bits
```

```
    crc = CRC4tab[crc] ^ (b & 0x0F) # Compute CRC for 4 most significant bits
# CRC for sequence number
# First 4 bits
crc = CRC4tab[crc] ^ (packed_frame[-2] >> 4)
# Last 8 bits of seq
crc = CRC4tab[crc] ^ (packed_frame[-1] >> 4)
crc = CRC4tab[crc] ^ (packed_frame[-1] & 0x0F)
# CRC value calculated
crc = CRC4tab[crc]
# Get the second-to-last byte and get only the 4 least significant bits
# packed_frame[-2] = s3 s2 s1 s0 c3 c2 c1 c0
# 0x0F = 0 0 0 0 1 1 1 1
# & is the bitwise and operator. For a bit x: x & 0 = 0; x & 1 = x.
frame_crc = (packed_frame[-2] & 0x0F) # frame_crc = 0 0 0 0 c3 c2 c1 c0
# Compare CRC value with the one sent in the packed frame
if crc == frame_crc:
    # CRC is valid
else:
    # CRC is invalid, either packed_frame is corrupted or frames do not have the expected format
```

3 Application Example

The `sense.py` sample application included in the ScientISST SENSE Python API repository[1] serves not only as an example of the capabilities of the API and how to use it, but also as a regular-use tool to acquire data from a CORE hardware device. It is intended to be used as a Command Line Interface (CLI) tool, using Python to make an acquisition with a desired setup, save the data on a human-readable `.csv` file, or make a real-time plot of the data. Figure 4 exemplifies a `.csv` file output from `sense.py` and Table 1 describes the key-value pairs present in the header of such files.

Having an example application with supporting documentation helps users quickly understand the functionality the API provides and how to implement it without having to understand the underlying code or needing assistance. To make an acquisition, the user can simply run the script using Python, while specifying the device's Bluetooth MAC address (or port number when using Wi-Fi modes) as an input argument.

[1] https://github.com/scientisst/scientisst-sense-api-python/tree/main

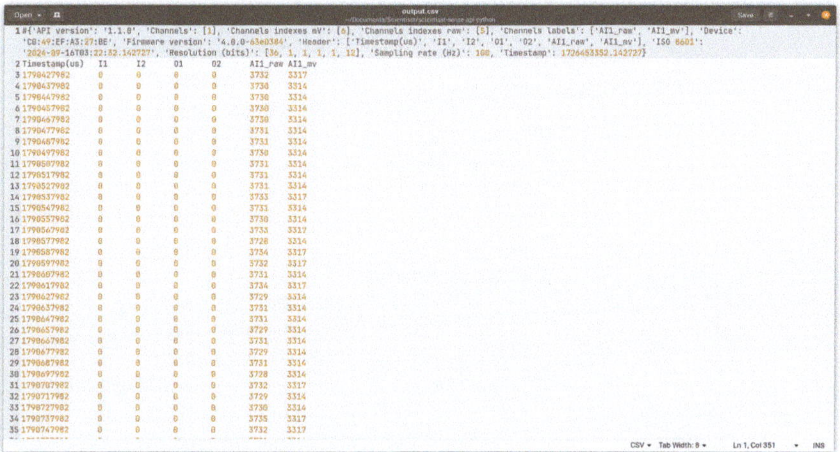

```
#{'API version': '1.1.0', 'Channels': [1], 'Channels indexes mV': [6], 'Channels indexes raw': [5], 'Channels labels': ['AI1_raw', 'AI1_mv'], 'Device':
'C8:49:EF:A3:27:BE', 'Firmware version': '4.0.0-63e0384', 'Header': ['Timestamp(us)', 'I1', 'I2', 'O1', 'O2', 'AI1_raw', 'AI1_mv'], 'ISO 8601':
'2024-09-16T03:22:32.142727', 'Resolution (bits)': [36, 1, 1, 1, 1, 12], 'Sampling rate (Hz)': 100, 'Timestamp': 1726453352.142727}
Timestamp(us)   I1   I2   O1   O2   AI1_raw AI1_mv
1790427982      0    0    0    0    3732    3317
1790437982      0    0    0    0    3730    3314
1790447982      0    0    0    0    3730    3314
1790457982      0    0    0    0    3730    3314
1790467982      0    0    0    0    3730    3314
1790477982      0    0    0    0    3731    3314
1790487982      0    0    0    0    3731    3314
1790497982      0    0    0    0    3730    3314
1790507982      0    0    0    0    3731    3314
1790517982      0    0    0    0    3731    3314
1790527982      0    0    0    0    3731    3314
1790537982      0    0    0    0    3733    3317
1790547982      0    0    0    0    3731    3314
1790557982      0    0    0    0    3730    3314
1790567982      0    0    0    0    3733    3317
1790577982      0    0    0    0    3728    3314
1790587982      0    0    0    0    3734    3317
1790597982      0    0    0    0    3732    3317
1790607982      0    0    0    0    3731    3314
1790617982      0    0    0    0    3734    3317
1790627982      0    0    0    0    3729    3314
1790637982      0    0    0    0    3731    3314
1790647982      0    0    0    0    3731    3314
1790657982      0    0    0    0    3729    3314
1790667982      0    0    0    0    3731    3314
1790677982      0    0    0    0    3729    3314
1790687982      0    0    0    0    3731    3314
1790697982      0    0    0    0    3728    3314
1790707982      0    0    0    0    3732    3317
1790717982      0    0    0    0    3729    3314
1790727982      0    0    0    0    3730    3314
1790737982      0    0    0    0    3735    3317
1790747982      0    0    0    0    3732    3317
```

Fig. 4 Example of a `.csv` output from `sense.py` using API mode ScientisstV2

```
# For omitted parameters, sense.py uses the following defaults:
# channels=1,2,3,4,5,6; sampling_rate=100; duration=unlimited,
↪ no output
$ python3 sense.py -c 1 08:3A:F2:49:AB:DE # Stream channel A1
$ python3 sense.py -c 1 -f 1000 08:3A:F2:49:AB:DE # Stream
↪ channel A1 at 1000Hz
$ python3 sense.py -c 1,2,4 -f 1000 08:3A:F2:49:AB:DE # Stream
↪ channels 1,2 and 4 at 1000Hz
```

The `-o [filename.csv]` option specifies that the data should be saved on a `.csv` file.

```
$ python3 sense.py -o output.csv 08:3A:F2:49:AB:DE # Save data to a
↪ file named output.csv
$ python3 sense.py -o output.csv -d 10 08:3A:F2:49:AB:DE # Acquire
↪ for 10 seconds and save data to a file named output.csv
```

Finally, the `-s` option enables the streaming of data using the Lab Streaming Layer (LSL) protocol [11] and visualizes it on a real-time plot (Fig. 5).

```
$ python3 sense.py -s 08:3A:F2:49:AB:DE # The following snippet will
↪ start streaming with the default configuration using LSL
# On another terminal, plot the real-time data
$ python3 -m pylsl.examples.ReceiveAndPlot
```

Table 1 Description of key-value pairs present in the header of the output file

Key	Description
API version	Internal API version
Channels	Array of integers of the number IDs of active channels (excluding digital inputs and outputs)
Channels indexes mV	Array of integers of column indexes of the `.csv` containing the values converted to mV (excluding digital inputs and outputs)
Channels indexes raw	Array of integers of column indexes of the `.csv` containing the raw ADC values (excluding digital inputs and outputs)
Channels labels	Array of strings of the labels of the active channels (excluding digital inputs and outputs)
Device	MAC address string of CORE hardware device
Firmware version	Firmware version string of the CORE hardware device
Header	Array of strings of the labels of the `.csv` columns
ISO 8601	Date and time of the start of the acquisition following the ISO 8601 standard
Resolution (bits)	Array of integers of the number of bits used to represent each of the values
Sampling rate (Hz)	Expected sampling rate in Hz of the acquisition
Timestamp	Date and time of the start of the acquisition in seconds since epoch

4 Summary

The ScientISST SENSE Python API simplifies interacting with ScientISST CORE devices for physiological signal acquisition. It provides an interface that abstracts away the complexities of the underlying electronics and communication protocols. This chapter provided an overview of the design goals of the API and explained key implementations of the interactions with CORE devices. The following key points summarize the contents discussed throughout the chapter:

- The API should, as much as possible, abstract the complexity of the underlying systems.
- The API should have multiple modes so as not to break compatibility with older devices or firmware versions when it is updated.

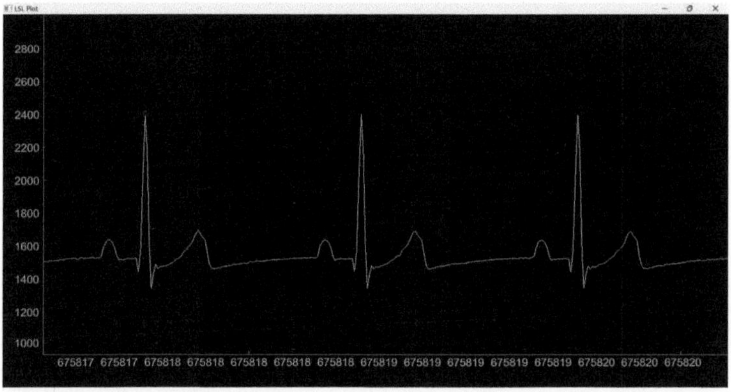

Fig. 5 Example of a LSL real-time plot of an electrocardiogram signal. Time in seconds on the x-axis and raw ADC values on the y-axis

- Efficient communication between the device and API is critical to ensure good performance.
- Bit packing is a vital and lightweight compression method when communicating with embedded devices such as CORE.
- Providing both the firmware and API allows implementing optimizations (such as a custom frame structure) that would not be possible if the CORE devices had to adhere to a more generic implementation.
- The system should be resilient to errors; per-frame CRC is one of the methods used to ensure that the frames transmitted by the CORE device are the same as those received by the API.
- Including an application example in the repository with adequate documentation provides a "hands-on" tutorial on how to use the API while also giving the API maintainer the perspective of the user; furthermore, it is a useful tool to quickly iterate on proof of concepts of use cases for the overall system.

Acknowledgments This work was partially funded by FCT/MECI through national funds and when applicable co-funded EU funds under project UID/50008: Instituto de Telecomunicações. It was also co-supported by the European Regional Development Fund (FEDER), through the Lisbon Regional Programme (LISBOA 2030) of the Portugal 2030 framework - LISBOA2030-FEDER-01318700 (ComSense).

References

1. F. Al-Turjman, *Internet of Nano-things and Wireless Body Area Networks (WBAN)* (CRC Press/Taylor & Francis Group, 2019)
2. F. Xia, A. Rahim, *MAC Protocols for Cyber-Physical Systems*. SpringerBriefs in Computer Science (Springer, Berlin, Heidelberg, 2015)

3. B. Hartpence, *Packet Guide to Core Network Protocols*. O'Reilly Series (O'Reilly Media, 2011)
4. J. Guerreiro, A. Lourenço, H. P. da Silva, A. Fred, Performance comparison of low-cost hardware platforms targeting physiological computing applications. Procedia Technol. **17**, 399–406 (2014)
5. H.P. da Silva, A. Lourenço, A. Fred, R. Martins, BIT: biosignal Igniter Toolkit. Comput. Methods Programs Biomed. **115**(1), 20–32 (2014)
6. E. Matthes, *Python Crash Course, 3rd Edition: A Hands-On, Project-Based Introduction to Programming* (No Starch Press, 2023)
7. E. Jones, T. Oliphant, P. Peterson, SciPy: Open Source Scientific Tools for Python—ScienceOpen. https://www.scienceopen.com/document?vid=ab12905a-8a5b-43d8-a2bb-defc771410b9
8. F. Pedregosa, G. Varoquaux, A. Gramfort, V. Michel, B. Thirion, O. Grisel, M. Blondel, P. Prettenhofer, R. Weiss, V. Dubourg, J. Vanderplas, A. Passos, D. Cournapeau, M. Brucher, M. Perrot, E. Duchesnay, G. Louppe, Scikit-learn: machine learning in Python. J. Mach. Learn. Res. **12**, 01 (2012)
9. M. Banahan, M. Santifaller, D. Brady, *The C Book: Featuring the ANSI C Standard* (Addison Wesley, Reading, 1991)
10. ITU-T, International Telecommunication Union ITU-T G.704 telecommunication standardization sector of ITU (10/98) series G: transmission systems and media, digital systems and networks digital transmission systems—terminal equipments—general synchronous frame structures used at 1544, 6312, 2048, 8448 and 44 736 kbit/s hierarchical levels (Recommendation G.704, International Telecommunication Union, 1998)
11. C. Kothe, S.Y. Shirazi, T. Stenner, D. Medine, C. Boulay, M.I. Grivich, T. Mullen, A. Delorme, S. Makeig, The lab streaming layer for synchronized multimodal recording (2024). bioRxiv

BioSPPy: Opening Physiological Signal Processing to the World

Patrícia Justo Bota, Rafael Silva, Sofia Monteiro, Carlos Carreiras, Ana Sofia Cacais do Carmo, and Hugo Plácido da Silva

1 Introduction

Physiological signals are measurements obtained from the human body or resulting from its activity. Also known as biosignals, physiological data are measures of physiological systems such as the cardiac or respiration cycle and are measured through physiological sensors. Alternatively, biosignals can result from the user interaction with their environment such as the speed and accuracy of a user typing on their computer keyboard. As a result, these signals can be categorized into several types:

1. **Bioelectrical**: Signals generated by electrical activity in biological tissues, such as action potentials or EEG signals
2. **Biomechanical**: Signals representing physical forces or displacements in biological systems, such as respiratory cycles or joint angles
3. **Biochemical**: Signals derived from the concentration or activity of biomolecules, such as hormone levels or neurotransmitter concentrations
4. **Biophysical**: Signals based on the physical properties of biological structures, such as optical signals or spectroscopic responses

In physiological computing, the most accessible body signals are related to the nervous, musculoskeletal, circulatory, respiratory, and integumentary systems, typically involving biomechanical or bioelectrical signals, such as Blood Volume

P. J. Bota (✉) · R. Silva · S. Monteiro · A. S. C. do Carmo · H. P. da Silva
Instituto Superior Técnico, Universidade de Lisboa, Lisbon, Portugal

IT - Instituto de Telecomunicações, Lisbon, Portugal

C. Carreiras
CardioID - Technologies, Lda., Porto, Portugal

H. P. da Silva et al. (eds.), *Open Source Biomedical Engineering*,
https://doi.org/10.1007/978-3-032-03655-1_4

Pulse (BVP), Electrocardiography (ECG), Electrodermal activity (EDA), and many others.

However, the available data is often in a raw format, requiring the application of signal processing algorithms for post-processing tasks such as noise filtering and extraction of meaningful information. Currently, this is performed either through closed-source proprietary tools or through open-source tools. Commercial solutions are validated and well documented and offer good support, although at the expense of elevated costs, lack of flexibility, and unknown processing pipelines. This has led researchers to create their own code, leading to the replication of diverse and redundant libraries, designed for a specific signal or application, improperly validated, and with high learning curves due to being used by field experts, with complex functionalities.

The lack of a validated, open-source, and centralized toolbox tailored to the specificities of physiological signals in a user-friendly design for widespread use is a bottleneck to making use of the available data for further developments in the field.

In this chapter, we demonstrate the creation of an open-source Python toolbox for physiological data processing and analysis—the BioSPPy library. BioSPPy bundles useful methods for 9 multiple commonly used physiological signal modalities (Electrocardiography (ECG), Electrodermal Activity (EDA), Heart Rate Variability (HRV), Electromyography (EMG), Photoplethysmography (PPG), Electroencephalography (EEG), respiration, Phonocardiography (PCG), accelerometer), in a structured and user-friendly design, integrating 7 modules from data loading to processing and knowledge extraction. BioSPPy was the first open-source Python library for the processing of physiological signals (according to initial commit in GitHub,[1] accessed on 19/04/2025).

By being open source, BioSPPy is free to use and allows for collaborative development and improvement, to help shape the library's structure, fix bugs, and add features, evolving faster than proprietary methods due to community contributions. Moreover, BioSPPy offers flexibility and extensibility, allowing users to adapt the tool to their particular goals and needs. Lastly, BioSPPy is transparent, allowing users to understand the algorithms and methods used in the processing of their data, linking to the peer-reviewed results obtained.

[1] https://github.com/PIA-Group/BioSPPy/commit/3015da11f5c972205356111e1759f6a2d2d029d8

2 Background

2.1 *What Languages Are Available for Physiological Computing Applications?*

The first decision that needs to be addressed for the implementation of a physiological computing system is the selection of its language.

Historically, Matlab [1], LabView [2], and R [3] have been the primary platforms for engineering and scientific computing in general, gaining a significant presence in biomedical engineering as the field unfolded. Although Matlab and LabView provide quite comprehensive and powerful environments, their practical application is often hindered by the high costs. Octave and R, on the other hand, while being open source, have remained somewhat specialized environments, for users with deeper technical backgrounds. Within this landscape, Python has seen significant growth over the recent years, reaching a leadership status in multiple domains.[2]

Matlab is widely established in academia, with a greater number of available toolboxes for physiological signal processing. Examples include the TEAP [4], Bio-SP tool [5], PRANA[6], EEGLAB [7]), AuBT [8], BioSigKit [9], and LivBioSig [10], just to name a few. However, Matlab is a paid service, limiting its usability. As an alternative, LabView is a free graphical programming environment, with available tools for physiological signal processing such as the LabView Biomedical Toolkit [11]. However, LabView has a complex integration with third-party environments and limited multiplatform support (especially for mobile operating systems).

In the latter years, the field has grown towards open source through languages such as Python or R. The R language is widely adopted in the statistics field with over 10,000 CRAN packages; available[3] however, it is considered slow, lacks support for dynamic or 3D graphics, and is a complicated language, with a higher learning curve for a beginner. The Python language has grown as an alternative, given its readability and speed, and is nowadays the leading language in machine learning and data science.

2.2 *Why Python?*

Python was created by Guido van Rossum in the late 1980s. It is an open-source interpreted, high-level, general-purpose programming language that allows procedural, object-oriented, and functional programming. One of its main design principles is code readability, being a dynamically typed and garbage-collected language, with no need to explicitly allocate and release memory for variables.

[2] https://spectrum.ieee.org/the-2018-top-programming-languages

[3] cran.r-project.org/web/packages/available_packages_by_name.html

Fig. 1 Typical pipeline for physiological data processing

Additionally, Python is supported in most chip architectures and operating systems. These qualities make Python one of the most popular programming languages, being the 1st most used language following IEEE spectrum.[4] For the field of physiological computing, Python is of special interest since open-source libraries such as `SciPy` [12] and `scikit-learn` [13] are available to perform low-level functions such as handling and processing of numerical data, from which physiological computing applications can be built.

3 Related Work

Figure 1 illustrates a typical pipeline for physiological data processing. It usually follows four main steps: (1) load data, through which the data in a file is loaded into the programming environment and stored in a variable; (2) filtering, where a low-pass, band-pass, or high-pass filter is applied to remove noise; (3) feature extraction, through which characteristics are extracted from the data to enable a compact and more informative characterization of the data; and (4) outlier removal, where data samples considered as outliers are removed.

Additional steps can be performed depending on the data, such as segmentation, interpolation, filling missing data, or data augmentation. Knowledge extraction (e.g., classification) was considered out of the scope since they are very dependent on the problem, and the libraries specific to this task such as TensorFlow [14] or scikit-learn [13] are developed and considered the state of the art tools. Moreover, these tasks are often performed based on features, independently of the input type (e.g., image, signals) from which the features were extracted, thus, not related to physiological signals.

A diverse set of physiological computing libraries available in the literature already seek to implement the processing steps illustrated in Fig. 1 for higher-level usage. Looking at the date of the first commit on GitHub, BioSPPy is one of the first ever created for physiological computing (in 2015) and has since then been used as a base for new platforms, such as NeuroKit2 [15] or pyHRV [16].

BioSPPy is one of the most comprehensive tools available in the literature, covering the majority of the available physiological signals in the literature, except for the Electrooculography (EOG) signal, as well as the required pipelines to process a physiological signal from its raw form to a feature vector. BioSPPy is also the only library that provides a module for feature extraction across four domains (time,

[4] https://spectrum.ieee.org/the-top-programming-languages-2023, accessed on May 14, 2024

frequency, phase space, and cepstral). The other libraries are more specialized, focusing on a specific signal or application and are not as comprehensive as BioSPPy. Besides its comprehensive nature, BioSPPy stands out by placing a special emphasis on the semantic nature of the function's inputs/outputs, so the functions and their results are easier to understand for all user levels. This is achieved by using keyword-based function input parameters and return values that improve their readability while still enabling native functionality. Additionally, BioSPPy provides a module with higher-level functions that allow working on the semantic keyword-based functions. This is a unique feature of BioSPPy, not present in the other libraries, except for NeuroKit2 [15].

4 Design Principles

4.1 Overview

To make BioSPPy simple and intuitive for the end user, a modular approach was chosen for its structure, focusing on the different biosignal processing tasks. This means that the set of operations within a specific domain is contained in one place. For example, because data visualization crosses most application domains, there is a module (i.e., a written Python script) denoted as `plotting.py` with all the variables and plotting functions of BioSPPy. This way of structuring a package comes with many advantages, namely:

- **Reusability:** Each module can access the functions of other modules, which avoids duplication of code.
- **Maintainability:** Modifications in one module are unlikely to disrupt the functionality of other modules.
- **Easy refactoring:** If a change of structure is needed, the modules can be easily rearranged.
- **Collaboration-friendly:** New functionalities can be added without major conflicts.

A standard Python package consists of different modules that can be organized into subpackages. BioSPPy (`bp`) follows the same structure, including subpackages with modules for each type of signal (e.g., `bp.signals`), a subpackage for feature extraction organized by the domain (`bp.features`) and modules for statistical tests (`bp.stats`), data visualization (`bp.plotting`), and others.

For specific recurrent tasks, BioSPPy integrates popular and useful third-party Python libraries such as `NumPy`, `Matplotlib`, and `SciPy`. This allows the user to, for example, read and write files, create plots, and compute metrics with a higher level of abstraction. To illustrate the advantages, instead of using the `NumPy`'s function `loadtxt` to read a text file (e.g., `.txt`, `.json`), the `bp.storage` module can be used, which can also automatically extract the metadata from the file, if available:

```
import biosppy as bp
signal, metadata = bp.storage.load_txt("./examples/pcg.txt")
sampling_rate = metadata["sampling_rate"]
```

In addition to *TXT*, there are also functions to read and write other file formats such as *JSON* and *HDF5*.

Moreover, the user can choose to process the biosignals with predefined parameters and easily obtain features of interest. Each biosignal module inside the `bp.signals` subpackage contains a main function with the module's name that automatically performs the basic signal processing steps to the input signal including filtering, segmentation, feature extraction, and visualization. As an example, to use the `pcg()` function from the `bp.signals.pcg` module, the user needs to input:

```
pcg_out = bp.signals.pcg.pcg(signal, sampling_rate)
```

4.2 *Structure*

As the last example detailed, the different functions and modules are accessed using the Python *dot notation*. A more detailed description of BioSPPy's structure, with a diagram of the current modules and subpackages, is presented in Fig. 2.

As it can be seen, the BioSPPy's subpackages and modules are organized according to different biosignal processing tasks,[5] making it easy for the user to access the desired information.

Another relevant aspect to take into account is the user experience with BioSPPy's output variables. Since the ultimate goal is to facilitate biosignal processing, the way the user interacts with the functions and modules was designed to be as simple and intuitive as possible, even for inexperienced Python users. This was the principle behind the creation of the object used to organize the outputs of the main functions: the `ReturnTuple`.

4.3 `ReturnTuple` *Object*

The `ReturnTuple` object is a hybrid form between a Python dictionary and a tuple. It is essentially a tuple with keys and values, which allows the user to interact with the output in different and versatile ways, including:

[5] The reader is encouraged to access BioSPPy's GitHub repository to discover all available tools for biosignal analysis and signal processing at https://github.com/scientisst/BioSPPy.

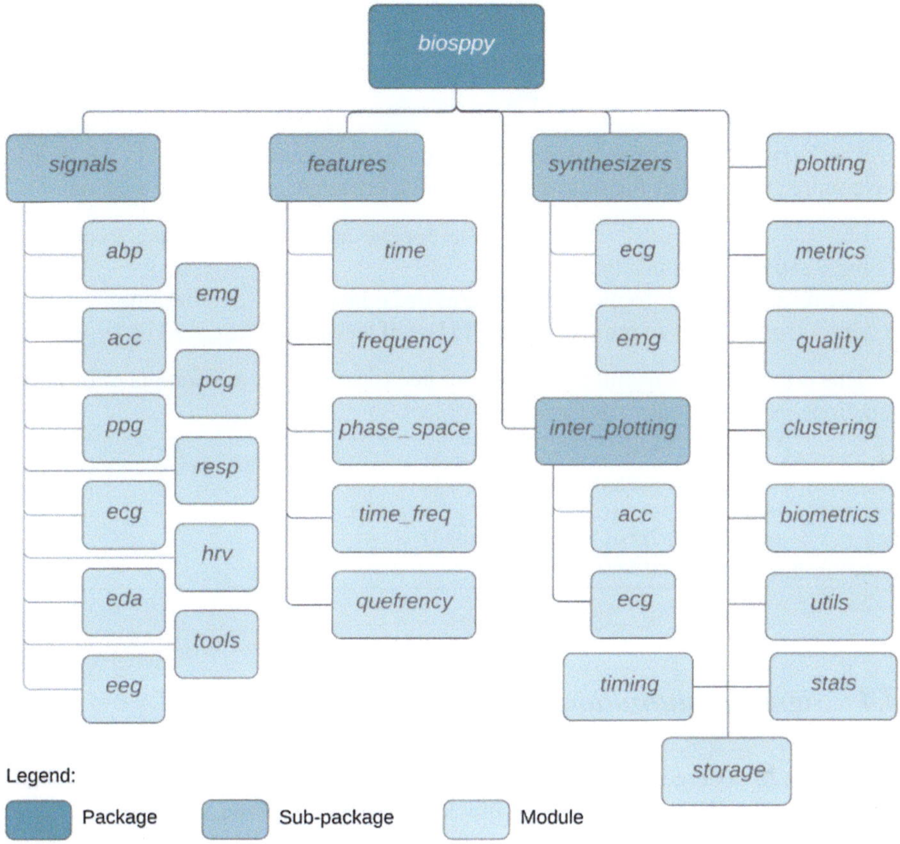

Fig. 2 BioSPPy's structure, including the package, subpackage, and module categories. Image extracted from [17]

– Accessing the list of value names inside the object

```
print(pcg_out.keys())
["ts", "filtered", "peaks", "heart_sounds", "heart_rate",
"systolic_time_interval", "heart_rate_ts", "inst_heart_rate"]
```

– Assigning variables individually with the correspondent name/key

```
pcg_filtered = pcg_out["filtered"]
```

– Unpacking the values from the functions, in their respective order

```
freqs, power = bp.signals.tools.power_spectrum(pcg_filtered)
```

– Converting to a dictionary using the method as_dict()

```
type(pcg_out.as_dict())
<class "collections.OrderedDict">
```

The ReturnTuple object is also equipped with methods that enable the user to have more control over the output format and its content, namely, the methods:

- append(new_values, new_keys) to append new keys and values to an existing ReturnTuple instance
- join(new_tuple) to merge a ReturnTuple instance to another one
- and delete(key) to remove a specific key of an existing ReturnTuple instance

This is particularly useful when the user needs to compile the outputs of different functions, or even create a personalized object.

```
s1_indices = pcg_out["heart_sounds"]==1
s1_peaks = pcg_out["peaks"][s1_indices]
pcg_out = pcg_out.append(s1_peaks, "s1_peaks")
pcg_out = pcg_out.delete("heart_sounds")
pcg_out = pcg_out.delete("peaks")
```

4.4 Signal Visualization

In addition to generating a ReturnTuple object, each "main" signal function calls the bp.plotting module allowing quick visualization of the signal and some of the extracted information. This functionality can be controlled using the optional Boolean parameter show. For example, when calling the bp.signals.pcg.pcg function we obtain the plot depicted in Fig. 3:

```
pcg_out = bp.signals.pcg.pcg(signal, sampling_rate, show=True)
```

As one can notice, the generated figure contains different plots with the information that the function bp.signals.pcg.pcg computes, including the raw signal, the filtered signal with the indication of the heart sounds, the instantaneous heart rate, and the classification of each heart sound into S1 or S2. This is achieved because a call to the bp.plotting.plot_pcg function is automatically made.

Since this figure is generated by a Matplotlib backend, the user can also interact with the different plots (e.g., zooming, panning) with the aid of the navigation toolbar. The plot functions also have an optional path parameter where the user can specify the filename to which the figure should be saved:

```
>>> pcg_out = bp.signals.pcg.pcg(signal, sampling_rate, show=True,
    path="pcg.png")
```

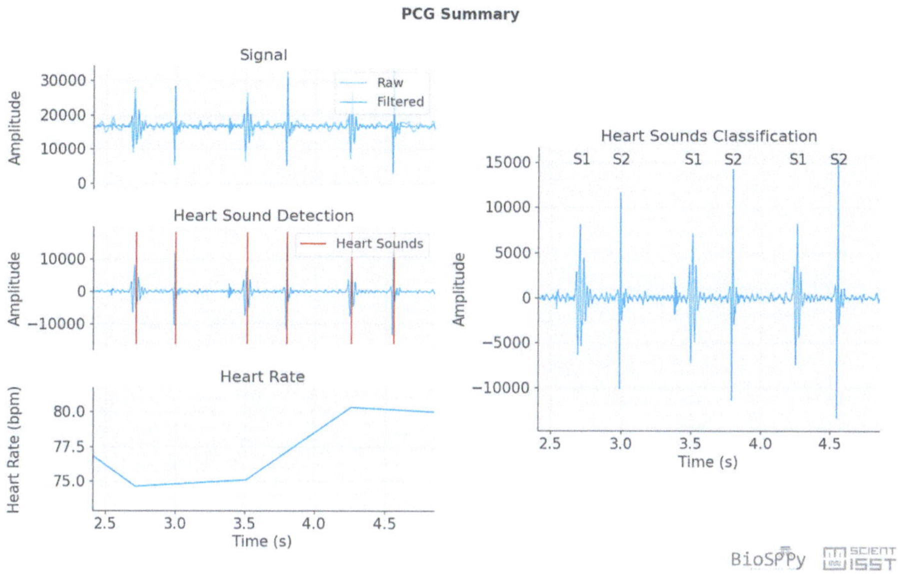

Fig. 3 Example of signal visualization in BioSPPy for a PCG recording

4.5 Package Distribution

BioSPPy follows the Semantic Versioning (SemVer) convention for assignment of package version numbers.[6] The SemVer number format is `MAJOR.MINOR.PATCH`:

- `MAJOR` increments indicate changes that might break backward compatibility. Example: updating from `1.2.4` to `2.0.0`
- `MINOR` increments indicate the addition of functionality in a backward-compatible way. Example: updating from `2.1.2` to `2.2.0`
- `PATCH` increments indicate small changes or bug fixes that do not break backward compatibility. Example: updating from `2.2.1` to `2.2.2`

Publishing of new package distribution releases to the Python Package Index (PyPI)[7] and the maintenance of a changelog for thorough documentation of changes are automated through GitHub Actions, ensuring a robust pipeline for package building and deployment.

BioSPPy also complies with the FAIR principles [18] (findable, accessible, interoperable, and reusable). The toolbox is hosted on GitHub, which ensures open access. BioSPPY was published under BSD 3-clause considering it guarantees copyright and requires the attribution of authorship, exempts the authors of the code

[6] https://semver.org/

[7] https://pypi.org/

from liability for improper use, allows the distribution of modified versions, and, lastly, allows for commercial use.

Additionally, BioSPPy is also available on Zenodo,[8] which facilitates proper versioning and citation of the software. Zenodo assigns a DOI (Digital Object Identifier) to each release, ensuring that users can cite specific versions of BioSPPy accurately, allowing for the reproducibility of experiments.

4.6 Documentation

BiosPPy's documentation is hosted at Read the Docs,[9] an open-source platform that allows automated documentation deployment. It provides version control, full-text search, and builds the documentation by pulling the doc files from the Github repository. It is integrated with Sphinx,[10] an automatic documentation generator that uses Python docstrings using NumPy's formatting and style.

Updates to the code are automatically reflected, making it easier to maintain accurate and up-to-date documentation. This way, we ensure that the documentation is both comprehensive and consistent, providing detailed descriptions of all subpackages, modules, and functions. Maintaining well-documented code is easier for others to understand, modify, and extend, facilitating new contributions and improvements to BioSPPy.

4.7 How to Contribute?

Contributing to BioSPPy can significantly benefit both the academic and industrial communities that work with biosignal processing. It also helps gather and share developments made by students and researchers, increasing their worldwide outreach and providing a referenceable resource. Contributions can have different impacts on the repository: they can add or change lines of code in a function (e.g., optimize an algorithm, add a new parameter to a function), a new function inside a module (e.g., a new algorithm to detect peaks in the ECG module), or an entirely new module made up of new functionalities (e.g., adding functions related to a new biosignal inside the `bp.signals` subpackage). Given the modular and biosignal-oriented structure of the toolbox, contributing becomes straightforward. However, because of the collaborative nature of this project, it is essential to adhere to the established structure and guidelines.

[8] https://zenodo.org/records/14900336

[9] https://about.readthedocs.com/

[10] https://www.sphinx-doc.org/en/master/

- **Respecting the structure**: New functionalities should be placed in the intended subpackage or module. For example, biosignal-specific functions should be added to the corresponding biosignal module, and more general functions for processing should be placed in the `bp.signals.tools` module.
- **Code consistency**: The contributor should follow the existing code style and format. This includes the code layout, the spacing, the name of the parameters, the documentation and references (docstrings), and the use of `ReturnTuple` as an output variable, among other considerations.
- **Commenting**: A short description should be provided for each step, and complex operations should be explained. On the other hand, over-commenting can make code harder to read and understand.
- **Reusing available functions**: Some signal processing tools are already implemented (e.g., filtering, feature extraction) and should be reused to avoid code duplication or unnecessary calls to third-party packages. Also, adding new dependencies should be avoided unless absolutely necessary.
- **Testing**: Before contributing, the code should be tested to ensure the stability of the toolbox and avoid breaking the functioning of other functions and modules. A working example should also be included for the reviewer to test. Unit testing for BioSPPy is currently under development.

The structure of the code should ensure compatibility with the API documentation. Each function's docstring has a description that explains its operations and references the source for the original function or algorithm. This is followed by a list of the function's input parameters and output parameters and what it returns, which should include the type of data (array, integer, float, etc.). Whenever possible, required function arguments should have a default value assigned with `keyword_name=value`. An example is presented in Fig. 4

In order to contribute, users are encouraged to get acquainted with the basic functioning of Git and GitHub and check the contribution guidelines available on BioSPPy's GitHub page. After submitting a pull request, reviewers may interact with the contributor to request adjustments. GitHub issues can also be opened to report bugs and malfunctions and make suggestions. Each contribution to BioSPPy enhances its functionality, promotes collaboration, and advances biomedical research and accessibility.

5 Application Examples

To illustrate the BioSPPy capabilities, we will show how to process the PPG signal, from its raw format to a cleaned feature representation.

A starting point is to load the data. The BioSPPy storage module provides a function to load the data from a text file. The data used in this example is available in the BioSPPy examples folder. The sampling rate is also extracted from the metadata.

```
def example_function(signal=None, sampling_rate=1000, show=False):
    """Example function. Calculate [feature] for signal [signal_name].
    Originally developed by [AuthorYear]_.

    Parameters
    ----------
    signal : array
        Input signal.
    sampling_rate : int, float, optional
        Sampling frequency (Hz).
    show : bool
        If True, plot the results.

    Returns
    -------
    feat1 : float
        Extracted feature 1.
    feat2 : array
        Extracted feature 2 array.

    References
    ----------
    .. [AuthorYear] Author 1, Author 2, et al. "Title"
    Journal/Repository, Year. Link

    """

    # compute features
    feat1 = ...
    feat2 = ...

    # show
    if show:
        plot_results(signal=signal, sampling_rate=sampling_rate, feat=feat2)

    args = (feat1, feat2)
    names = ('feature1_name', 'feature2_name')

    return utils.ReturnTuple(args, names)
```

Name of the function
Snake case convention
Parameters
Consistent names and default values specified
Description of the function
Variables definition
Name, type and description
DOCSTRINGS
Numpydoc style
Enclosed with triple quotes
Sections for parameters, outputs, references and notes
Output variables
Use of *ReturnTuple* object for named variables

Fig. 4 Example function layout and formatting guidelines used in BioSPPy's code

Listing 1 Load the data using the BioSPPy storage module

```
ppg_raw, ppg_metadata = biosppy.storage.load_txt("ppg.txt")
sampling_rate = ppg_metadata["sampling_rate"]
```

The resulting raw PPG signal is shown in Fig. 5a. The figure shows a noisy PPG, to get a better understanding of the noise present in the signal, we can plot the power spectrum of the signal (Fig. 5b). The power spectrum is calculated using the `bp.signals.tools.power_spectrum` function:

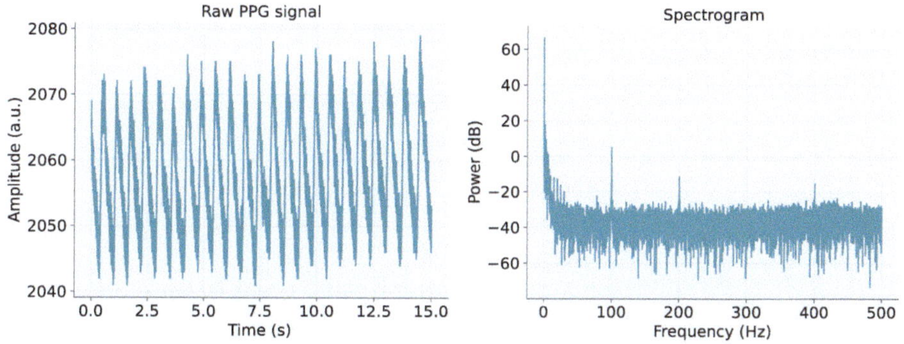

Fig. 5 Example of PPG data and its power spectrum upon processing with BioSPPy. (**a**) Example of PPG data available at BioSPPy examples folder. (**b**) Power spectrum of the PPG signal

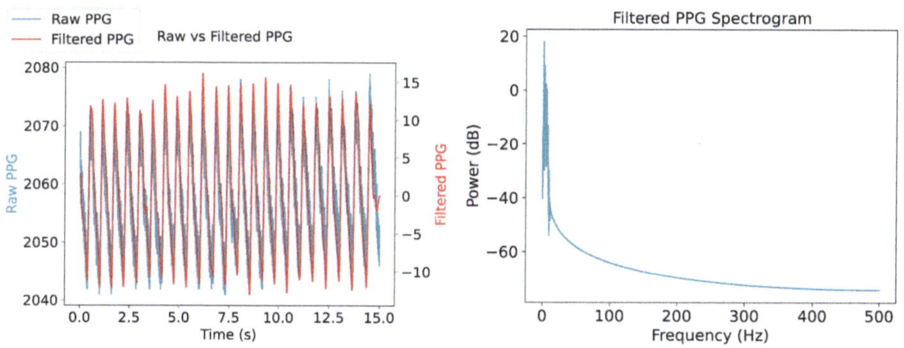

Fig. 6 Comparison of PPG data and its power spectrum. (**a**) Example of PPG data available at BioSPPy examples folder. (**b**) Power spectrum of the PPG signal

```
from biosppy.signals.tools import power_spectrum
freq, power = power_spectrum(ppg_signal, sampling_rate)
```

The power spectrum shows a higher magnitude for lower values and a peak at the heart rate frequency (around 100 Hz), which is the main frequency of interest in the PPG signal, and its harmonics. To remove the noise from the signal, we can use a bandpass filter. The BioSPPy filtering module provides a function to filter the signal using a Butterworth bandpass filter between 1 and 8 Hz. The filter is applied to the signal using the `bp.signals.tools.filter_signal` function. The filtered signal and its spectrogram are shown in Fig. 6.

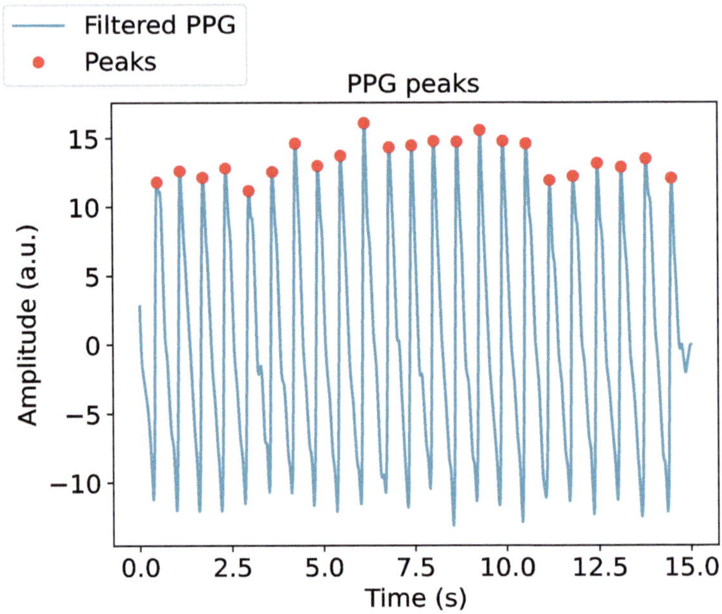

Fig. 7 Peaks extracted from the PPG signal

```
import biosppy.signals.tools as st
filtered_ppg, _, _ = st.filter_signal(signal=ppg_signal,
                                      ftype='butter',
                                      band='bandpass',
                                      order=4,
                                      frequency=[1, 8],
                                      sampling_rate=sampling_rate)
```

From the filtered signal, it is possible to extract the PPG features. The BioSPPy `bp.signals.ppg` module provides a function to extract the PPG pulse peaks. The function returns a dictionary with the extracted peak indices (Fig. 7).

```
pulse_peaks, _ = ppg.find_onsets_elgendi2013(signal=filtered_ppg,
                                             sampling_rate=sampling_rate)
```

The BioSPPy library provides a set of functions to analyze the quality of the signal. Namely, the `cSQI` function calculated using the `bp.signals.quality.cSQI` function, which returns a dictionary with the calculated signal quality values.

```
import biosppy.signals.quality as quality
cSQI = quality.cSQI(rpeaks=pulse_peaks)
```

From the pulse peaks, we can calculate the heart rate. The BioSPPy get_heart_rate module provides a function to calculate the heart rate from the pulse peaks. The function returns a dictionary with the heart rate values.

```
    import biosppy.signals.tools as st
hr_idx, hr = st.get_heart_rate(beats=pulse_peaks,
      sampling_rate=sampling_rate,
      smooth=True,
      size=3)
```

In this example, the signal quality index returned a score of 0.19, denoting an optimal signal. From the heart rate, a diverse set of features can be extracted to characterize the PPG signal. An example is the time-domain SDNN and RMSSD features. The SDNN and RMSSD features are calculated using the bp.signals.hrv.hrv_timedomain function. The function returns a dictionary with the calculated features. The SDNN and RMSSD features are extracted from the dictionary.

```
from biosppy.signals import hrv
rri = hrv.compute_rri(pulse_peaks, sampling_rate=sampling_rate)
out = hrv.hrv_timedomain(rri=rri)
SDNN, RMSSD = out['sdnn'], out['rmssd']
```

The SDNN denotes the standard deviation of successive pulse peak difference (also denoted as RR intervals) and shows the dispersion of the RRI values around its mean value. A higher variability is related to a higher sympathetic activity and is observed in health subjects, while a lower variability is observed in patients with heart disease.

The RMSSD obtains the root mean square of successive pulse peak difference (RRI). The RMSSD is related to the parasympathetic activity and is used to evaluate the vagal tone. A higher RMSSD value is related to a higher capability of adjusting one's heart rate and is observed in healthy subjects.

Moreover, the PPG signal can be decomposed into its frequency components, namely, the ultralow ($<$0.003 Hz), very low (0.003 to 0.04 Hz), low (0.04 to 0.15 Hz), high (0.15 to 0.40 Hz), and very high ($>$0.40 Hz) bands (Fig. 8).

To get a better understanding of the sympathetic/parasympathetic balance, we can calculate the LF/HF ratio [19]. The LF/HF ratio is calculated using the bp.signals.hrv.hrv_frequencydomain function. The function returns a dictionary with the calculated features. The LF/HF ratio is extracted from the dictionary.

```
from biosppy.signals import hrv
out = hrv.hrv_frequencydomain(rri=rri)
lf_hf = out['lf_hf']
```

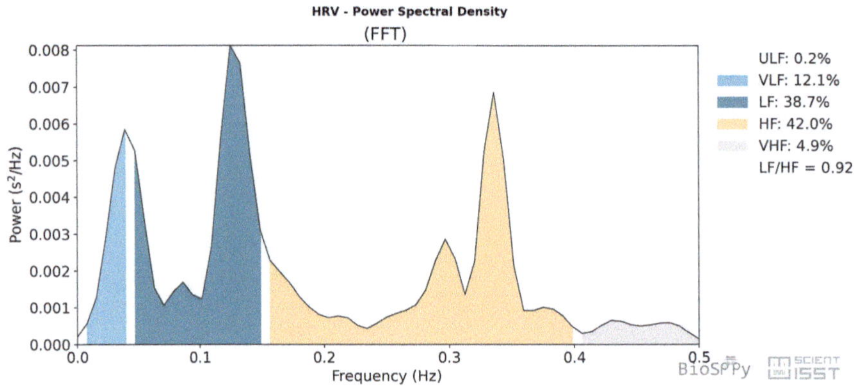

Fig. 8 Frequency domain decomposition of the interbeat intervals extracted from the PPG signal. Nomenclature: *ULF* ultralow frequencies, *VLF* very low frequencies, *LF* low frequencies, *HF* high frequencies, *VHF* very high frequencies

6 Summary

Wearable systems and physiological sensors are increasingly collecting vast data, highlighting the need for tools like BioSPPy. BioSPPy aids in processing and analyzing physiological signals, enhancing accessibility and innovation in the field. The following key points summarize the contents discussed throughout the chapter:

- Open-source software democratizes access to advanced tools and fosters collaboration for innovation.
- Open-source coding libraries provide low-cost, transparent, and flexible solutions for processing and analyzing physiological data.
- BioSPPy integrates all main steps in data processing, from data loading to feature extraction, including low-pass, band-pass, and high-pass filtering and the extraction of over 80 features.
- Structured semantically to make input/output and function interconnections clear and explainable.
- Open-source communities contribute to continuous improvement and adaptation of BioSPPy, with frequent updates.
- Developed with reusability, maintainability, easy refactoring, and collaboration in mind.

Acknowledgments This work was partially funded by FCT/MECI through national funds (FCT grants 2020.06675.BD and 2022.04901.CEECIND) and when applicable co-funded EU funds under project UID/50008: Instituto de Telecomunicações. It was also co-supported by the European Regional Development Fund (FEDER), through the Lisbon Regional Programme (LISBOA 2030) of the Portugal 2030 framework—LISBOA2030-FEDER-01318700 (ComSense)

References

1. MATLAB, *Multi-Paradigm Numerical Computing Environment* (The MathWorks, Natick, 2019)
2. C. Elliott, V. Vijayakumar, W. Zink, R. Hansen, National Instruments LabVIEW: a programming environment for laboratory automation and measurement. J. Assoc. Lab. Autom. **12**(1), 17–24 (2007)
3. R Core Team, *R: A Language and Environment for Statistical Computing* (R Foundation for Statistical Computing, Vienna, 2019)
4. M. Soleymani, F. Villaro-Dixon, T. Pun, G. Chanel, Toolbox for emotional feature extraction from physiological signals (TEAP). Front. Inform. Commun. Technol. **4**, 1 (2017)
5. M. Nabian, Y. Yin, J. Wormwood, K.S. Quigley, L.F. Barrett, S. Ostadabbas, An open-source feature extraction tool for the analysis of peripheral physiological data. IEEE J. Transl. Eng. Health Med. **6**, 1–11 (2018)
6. PRANA, *Polygraphic Recording Analyzer Software Suite* (PhiTools, Strasbourg, 2019)
7. A. Delorme, S. Makeig, EEGLAB: an open source toolbox for analysis of single-trial EEG dynamics including independent component analysis. J. Neurosci. Methods **134**(1), 9–21 (2004)
8. J. Wagner, J. Kim, E. Andre, From physiological signals to emotions: implementing and comparing selected methods for feature extraction and classification, in *Proceedings of the IEEE International Conference on Multimedia and Expo* (2005), pp. 940–943
9. T. Lorrain, I.K. Niazi, O. Thibergien, N. Jiang, D. Farina, LivBioSig: development of a toolbox for online bio-signals processing and experimentation, in *Proceedings of the Annual International Conference of the IEEE Engineering in Medicine and Biology Society (EMBC)* (2011), pp. 7302–7305
10. H. Sedghamiz, BioSigKit: a matlab toolbox and interface for analysis of biosignals software. J. Open Source Softw. **30**(3), 1–5 (2018)
11. Biomedical User Group, *LabVIEW for Biomedical Signal Analysis* (National Instruments, Austin, 2019)
12. P. Virtanen, R. Gommers, T.E. Oliphant, M. Haberland, T. Reddy, D. Cournapeau, E. Burovski, P. Peterson, W. Weckesser, J. Bright, S.J. van der Walt, M. Brett, J. Wilson, K.J. Millman, N. Mayorov, A.R.J. Nelson, E. Jones, R. Kern, E. Larson, C.J. Carey, İ. Polat, Y. Feng, E.W. Moore, J. VanderPlas, D. Laxalde, J. Perktold, R. Cimrman, I. Henriksen, E.A. Quintero, C.R. Harris, A.M. Archibald, A.H. Ribeiro, F. Pedregosa, P. van Mulbregt, SciPy 1.0 Contributors, SciPy 1.0: fundamental algorithms for scientific computing in Python. Nat. Methods **17**, 261–272 (2020)
13. F. Pedregosa, G. Varoquaux, A. Gramfort, V. Michel, B. Thirion, O. Grisel, M. Blondel, P. Prettenhofer, R. Weiss, V. Dubourg, J. Vanderplas, A. Passos, D. Cournapeau, M. Brucher, M. Perrot, E. Duchesnay, Scikit-learn: machine learning in Python. J. Mach. Learn. Res. **12**, 2825–2830 (2011)
14. M. Abadi, A. Agarwal, P. Barham, E. Brevdo, Z. Chen, C. Citro, G.S. Corrado, A. Davis, J. Dean, M. Devin, S. Ghemawat, I. Goodfellow, A. Harp, G. Irving, M. Isard, Y. Jia, R. Jozefowicz, L. Kaiser, M. Kudlur, J. Levenberg, D. Mané, R. Monga, S. Moore, D. Murray, C. Olah, M. Schuster, J. Shlens, B. Steiner, I. Sutskever, K. Talwar, P. Tucker, V. Vanhoucke, V. Vasudevan, F. Viégas, O. Vinyals, P. Warden, M. Wattenberg, M. Wicke, Y. Yu, X. Zheng, TensorFlow: lARGE-scale machine learning on heterogeneous systems (2015). Software available from tensorflow.org
15. D. Makowski, T. Pham, Z.J. Lau, J.C. Brammer, F. Lespinasse, H. Pham, C. Schölzel, S.H.A. Chen, NeuroKit2: a Python toolbox for neurophysiological signal processing. Behav. Res. Methods **53**, 1689–1696 (2021)
16. P. Gomes, P. Margaritoff, H. Silva, pyHRV: development and evaluation of an open-source Python toolbox for heart rate variability (HRV), in *Proceedings International Conference on Electrical, Electronic and Computing Engineering (IcETRAN)* (2019), pp. 822–828

17. P. Bota, R. Silva, C. Carreiras, A. Fred, H. P. da Silva, BioSPPy: a Python toolbox for physiological signal processing. SoftwareX **26**, 101712 (2024)
18. M.D. Wilkinson, M. Dumontier, I.J. Aalbersberg, G. Appleton, M. Axton, A. Baak, N. Blomberg, J.-W. Boiten, L.B. da Silva Santos, P.E. Bourne, et al., The fair guiding principles for scientific data management and stewardship. Sci. Data **3**(1), 1–9 (2016)
19. P.M. Caridade Gomes, *Development of an open-source Python toolbox for heart rate variability (HRV)*. Ph.D. Thesis, Hochschule für angewandte Wissenschaften Hamburg, 2019

Interactive Machine Learning: Strategies for Live Performance Using Electromyography

Michael Zbyszyński, Atau Tanaka, and Federico Visi

1 Introduction

Electromyography (EMG) from arm muscles has been used in the biomedical and human-computer interaction fields as a highly sensitive way to capture human movement and the intention of making gesture. EMG has been used in sign language recognition [1] and in creating discrete interactions [2, 3]. Commercial products in wearable computing have used EMG and Inertial Measurement Units (IMUs) in multimodal configurations as human interface devices [4]. EMG has been used as a signal with which to sense musical gesture [5–7]). Using EMG for music presents several challenges. The signal itself resembles noise and sensing such a low-voltage signal is difficult to do without accumulating noise from the environment. Individual anatomies vary (for example, the palmaris longus muscle is absent in about 14% of the population [8]) and we each employ our muscles differently, even when performing what looks like the same gesture. Basic signal processing can only go so far when interpreting expressive, nuanced biosignals.

We present methods for designing real-time, embodied [9] interactions using diverse sensors in conjunction with Interactive Machine Learning (IML) [10]. Specifically, we have employed EMG in conjunction with other sensors in the

---

M. Zbyszyński
L-Acoustics, London, UK

A. Tanaka (✉)
Department of Computing, Goldsmiths University of London, London, UK
e-mail: a.tanaka@gold.ac.uk

F. Visi
Luleå University of Technology, School of Music in Piteå, Piteå, Sweden

Universität der Künste Berlin, Berlin, Germany

Einstein Center Digital Future, Berlin, Germany

H. P. da Silva et al. (eds.), *Open Source Biomedical Engineering*,
https://doi.org/10.1007/978-3-032-03655-1_5

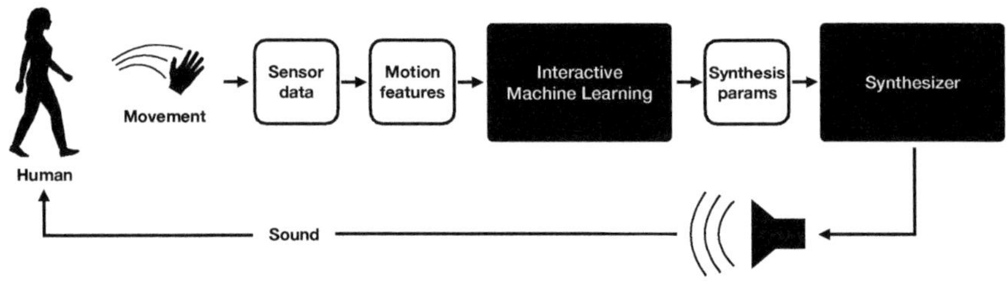

Fig. 1 Full signal flow for an interaction involving multiple sensors and interactive machine learning

creation of digital musical instruments that can be performed expressively with physical gesture.

IML offers the possibility to create bespoke interactions quickly and intuitively. Performers provide examples of a desired interaction to an algorithm which then generates a model that allows them to perform expressive variations of the provided examples as well as exploring new gesture-timbre relationships that are connected to the original ideas. IML functions both as an efficient way to generate interactions quickly and without the need for complex coding, as well as a creative tool for generating new sonic affordances that are tailored to a specific performers concept and physiology.

While the work presented here follows current practice in digital musical instrument design, the methods can be generalised to any situation demanding real-time, expressive control. We refer, in this chapter, to literature from biomedical research; control of prosthetic limbs is an important application of electromyography. These approaches may find broader application in examining the creation of rich media interaction for people with non-normative physical or mental abilities [11], as well as applications in rehabilitation, sports training, and product design [12].

An interaction signal flow situating EMG and IML between performer gesture and musical sound output is shown in Fig. 1.

This chapter is structured as follows. We begin by giving an overview of electromyography and how to capture EMG signals. Then we move on to extracting useful features from one or more EMG channels and use of signal processing on a single EMG channel to recognise, map, or track gestural data. For more complex interactions, involving multiple EMG channels in conjunction with other movement sensors, we explore the use of machine learning to identify gestures and poses, or to create regression models that smoothly interpolate between learned gestures. We finish by presenting a number of software resources for researchers and practitioners wishing to use the techniques presented here.

2 What Is Electromyography?

EMG is a measurement of the electrical activity produced by skeletal muscles. This electrical activity is correlated with our muscle generation of force and motion and allows for insight into physiological processes that we subconsciously employ to interact with the world. However, muscular activation and the resulting EMG signal are quite complex [13]. Muscle tension can be generated with no change in muscular length. This is called isometric activity, such as when pushing against a fixed object. Isotonic activity, on the other hand, involves changing the length of the muscle, such as when picking up an object [14].

One muscle (for example, a bicep) is made up of a number of individual motor units made up of a single motor neuron and a collection of muscle fibres. A nerve impulse, called an activation potential, is transmitted through the neuron to the muscle fibres, causing them to activate (if they have not already). These Motor Unit Action Potentials (MUAPs) can be measured by electrodes, and an EMG is typically made up of some of the multiple MUAPs from across the muscle. A sustained muscular contraction is caused by repetitive MUAPs.

The voltage of an EMG signal is relatively small. An individual MUAP has an average amplitude of around 100 μV, while a typical gesture involving multiple motor units has a unamplified range of ±5 mV. Muscles work like a "crowd-sourcing" model; the specific motor units that are recruited during any gesture is unpredictable and changes over the course of an action. There is also some muscular background noise, as motor units fire randomly. Because of this it is most effective to use statistical methods, as described below, to evaluate the current amplitude and level of activation represented by a given EMG sensor. Frequency components of an EMG signal can be analysed in the range from 0 to 500 Hz in healthy individuals, with most of the energy in the 50–150 Hz range.

EMG signals are strongly influenced by individual physiology. Similar motion can create very different signals when performed by different subjects, or by the same subject in different sessions. We have explored the question of "sameness" of gesture and proposed a non-Cartesian method of gesture design for different people creating similar movements tracked by EMG [15]. Here, this is one of the motivations for our use of interactive machine learning, as described below. By designing interactions on the fly, we are able to adjust to individual users.

3 Capturing Signal

3.1 Electrodes

The EMG signal can be measured in a noninvasive manner using surface-mounted electrodes (Fig. 2) (It is also possible to measure EMG using invasive techniques—needles or wires inserted directly into the muscle, but this procedure is not well

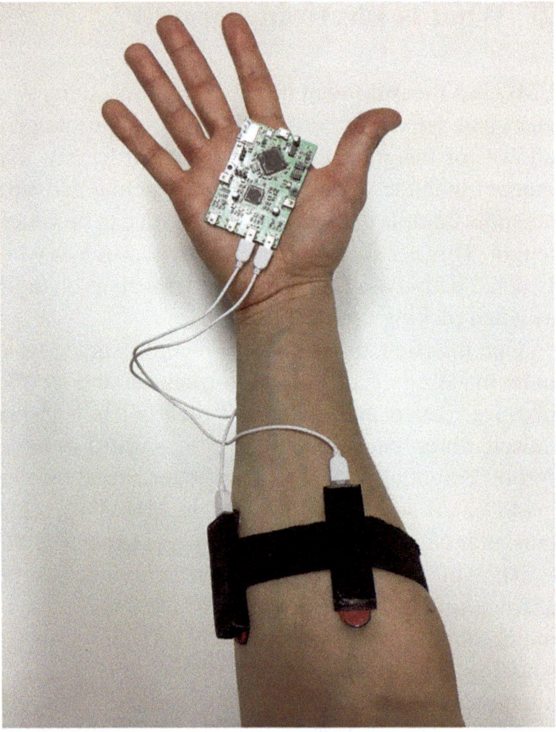

Fig. 2 EAVI EMG board with dry electrodes worn on the forearm

suited to DIY applications). Surface EMG electrodes make electrical contact with the surface of the skin. The best contact can be made with wet gel electrodes[1] made from silver/silver chloride. This requires some preparation of the skin and such electrodes have a limited lifespan. Dry electrodes[2] have the advantage in ease of application and reuse, although their conductivity is not equal to gel electrodes.

3.2 Electrode Placement

Electrodes are typically placed near the middle, "meaty" part of the muscle to be sensed [16]. Muscles in the forearm (Fig. 3) are a common place to put sensors, as these muscles control the hand and fingers. Other muscles in the arms, legs, back, or even face can be successfully sensed. This allows a wide range of applications beyond music, including performance and dance sensing, gaming, or designing interfaces for users with non-normative physical abilities, such as paralysis.

[1] https://www.pluxbiosignals.com/products/gelled-self-adhesive-disposable-ag-agcl-electrodes

[2] https://www.pluxbiosignals.com/products/non-gelled-reusable-ag-agcl-pack-10

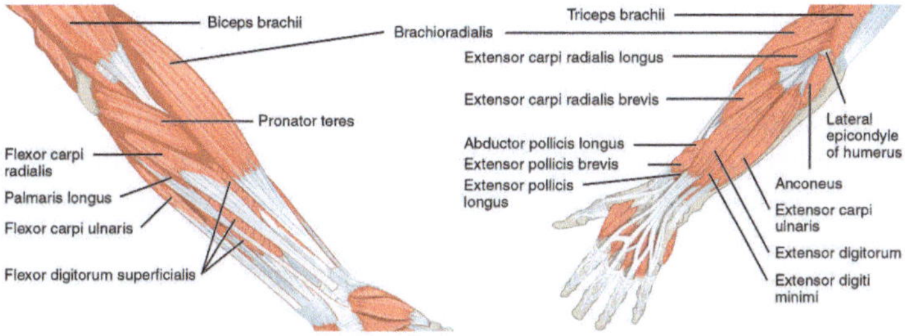

Fig. 3 An overview of the skeletal muscles that move the forearm (adapted from the following source: https://upload.wikimedia.org/wikipedia/commons/7/73/1120_Muscles_that_Move_the_Forearm.jpg)

MUAPs are measured by the difference between two electrodes placed at different sites on the same muscle. The signals are subtracted and the result is amplified; any signal at both electrodes is therefore eliminated. This circuitry is familiar to musicians from balanced signal cables used for microphones. A third electrode is used in an unrelated area, often joints or bony areas, to act as a reference ground that filters out ambient electrical noise. Electrodes sense activation of motor units which are close to the surface. Motor units may change position, relative to an electrode, as muscle length changes. In practice, this is less problematic than gross electrode shift on the surface of the skin. Reliable and consistent capture position is an important factor in reproducing interactions, especially with machine learning sessions where training data may have been recorded in a separate session to the performance.

3.3 Multiple Electrodes

The work we report on here focuses on the muscles in the forearm. Humans have a great deal of fine control on the motions of the hand and wrist, and this is a convenient place to capture expressive EMG data. Getting a complete picture of forearm muscular activity requires multiple EMG sensors. We have used configurations with four or eight sensors around the perimeter of the lower arm. There are four main muscle groups in the forearm, so such a configuration allows capturing the range of wrist motion with a sensor on or near each group. Redundant sensors can be helpful to deal with the fact that muscles move in relation to the electrodes, or to create a setup that accommodates users with different anatomies. A potential drawback comes from crosstalk between electrodes [17].

3.4 Sources of Noise

Noise is a substantial challenge when processing EMG signal. The signal itself strongly resembles noise. It is an aperiodic stochastic pulse train where the electrical potential is in the microvolt range. Furthermore, the low voltage of EMG signals makes them susceptible to electrical noise from the surrounding environment. The frequency of alternating current (60 Hz in the USA, 50 Hz in Europe and the UK) can be attenuated with a notch filter. The body itself acts as an antenna, as do any cables connecting sensors to the analogue to digital converter. Traditional best practice recommends amplifying the sensed voltage as close to the electrode is possible. This can be processed using differential amplification, so signal that is at both electrodes will be cancelled out. In our recent work, we have designed hardware to acquire and digitise unamplified differential signal at a sufficiently high resolution to apply gain and noise reduction in the digital domain [18].

4 Feature Extraction

Robust feature extraction is critical for deriving useful information from the data stream, given the complexity of the EMG signal. The signal, being a pulse train, has an intrinsically noisy characteristic, making differentiation from extraneous interference more difficult than in classical noise reduction scenarios. Signal from a single electrode is modelled well by Gaussian noise with an amplitude envelope that corresponds to the amount of tension in that muscle. Envelope following, a term borrowed from sound synthesis [19], can therefore be an effective first step in measuring gesture exertion.

Standard digital signal processing techniques can be employed to extract the amplitude envelope from the stream of EMG values. Taking the Root Mean Square (RMS) of a buffer of samples (50–100 milliseconds, depending on the sampling rate) is effective for this task. For N values, the RMS is:

$$x_{rms} = \sqrt{\frac{1}{N}\sum_{i=1}^{N} x_i^2}$$

Each sample is squared, the squares are averaged, and the square root of the result is taken. As this effectively smooths the signal across the number of samples in the buffer, there is a trade-off between stability and responsiveness. If the buffer is large (N is greater), then the envelope will be more smooth and stable. However, the RMS will respond more slowly to changes in tension (see Fig. 4. Smaller buffers respond more quickly, but give more jittery results. Latency (or perceived delay) will also increase as the buffer size increases.

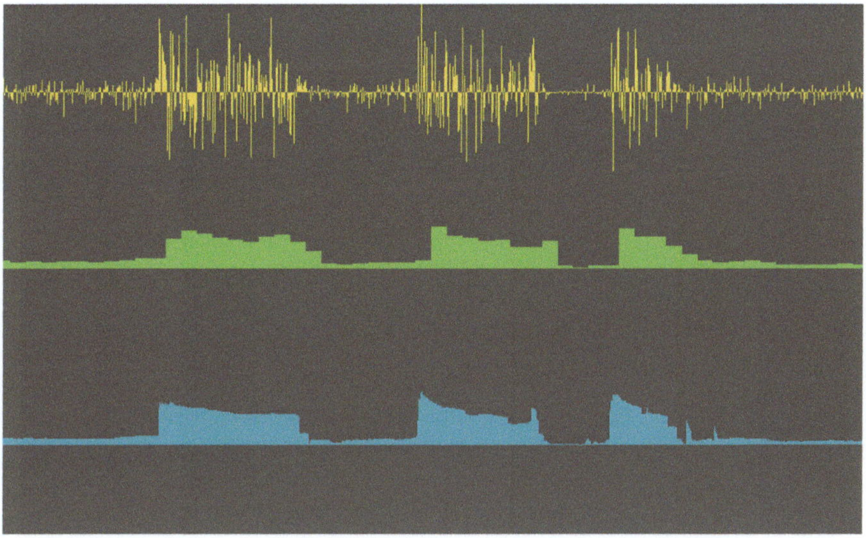

Fig. 4 Raw EMG signal (top) and amplitude envelope calculated by RMS (middle) and a Bayesian filter (bottom)

A more advanced technique that frees us from this trade-off uses probabilistic filtering. Recursive Bayesian filtering [20, 21] is used to estimate the current amplitude of the "neural drive", given a series of EMG readings. The Bayesian algorithm models the relationship between momentary readings and drive signal as a hidden Markov model (see Sect. 5.4) and is parameterised so that the smoothness and responsiveness of the output can be controlled independently. Because the filter is predictive, it can respond more quickly to abrupt changes in amplitude while maintaining a higher signal-to-noise ratio than RMS. Responsiveness is necessary for real-time musical applications where timing is crucial.

It is interesting to note that EMG tracks the signal that activates a muscle. That signal precedes actual motion of the joints by many milliseconds. In an application where we were interacting with a performer who was used to playing a button-based drum interface, we found that we were able to identify their triggering motion much sooner than they were expecting. The result was that drum samples were being triggered too early, and we had to add a short (50-ms) delay to our signal path to make the timing of interaction feel right to this performer. In this sense, the EMG is a gesture sensing modality that senses the intention of movement, compared to the resulting movement that is measured by external physical sensors like optical motion capture or IMUs [22].

The actual electrical activity of an EMG signal measured from an electrode attached to an individual can be affected by many factors, including the type of electrode, location relative to the muscle, amount of subcutaneous fat between the electrode and the muscle, the type of muscle, and so on. Because of this, it is usually useful to calibrate an EMG system to the voltage range measurable from a given electrode and user combination. It is common to take the amplitude envelope; measure the voltage correlated with the greatest tension a muscle can generate, called Maximum Voluntary Contraction (MVC); and express measurements as a percentage of that value. So, a mildly tensed muscle might be expressed as having an EMG amplitude of 20% of MVC.

In addition to amplitude, there are other features which are potentially useful [14]. The mean absolute value of samples in a window gives energy information. Waveform length (waveLen) is a measure of the complexity of the signal:

$$waveLen = \sum_{k=1}^{N} f(|x_k - x_{k+1}|)$$

Slope sign change is related to frequency and is the number of times the slope changes sign in an analysis window. Because EMG signal is aperiodic, frequency domain features typically do not perform well in gesture recognition tasks [23]. However, we have observed that mean frequency decreases with increased gesture execution speed. Spectral centroid increases with tension, but is not detected in overall spectral energy [24].

Additional features are possible with multiple EMG streams. With multiple channels of sensors placed around the forearm, it is possible to identify whether different muscles are acting in isometric opposition to each other. By isometric, we mean static contraction of muscles without any actual resulting limb movement. This is a specific affordance of the EMG as sensing modality—opposing muscles might not generate any limb movement that would be visible to other motion tracking systems such as optical motion capture or accelerometers and IMUs. The sum of all EMG amplitudes in our systems provides a measure of global exertion. We have developed a new compound feature called "vector sum" that reports amplitude and direction. Vector sum (Fig. 5) exploits limb physiology where muscles around the arm are situated in such a way that they can oppose or reinforce the action of other muscles. To calculate the vector sum in a case with eight sensors, we model each sensor as representing a vector pointing away from the centre of a circle, evenly spaced every 45°. The direction for each constituent sensor vector remains constant and the magnitude is proportional to the amplitude calculated by the Bayesian filter. The vectors are summed, giving a result that changes direction and magnitude related to the overall direction of force represented by all of the electrodes. When compared to the sum of all electrodes, the vector sum can distinguish gestures where muscles are opposing one another isometrically. This can be useful in cases where joint movement might be minimal, but the subjective perception of effort is quite high. The vector sum is reported as a pair of Cartesian coordinates (x_{vs}, y_{vs}), which are better suited to post-processing and subsequent machine learning than polar coordinates because they do not wrap around at zero degrees.

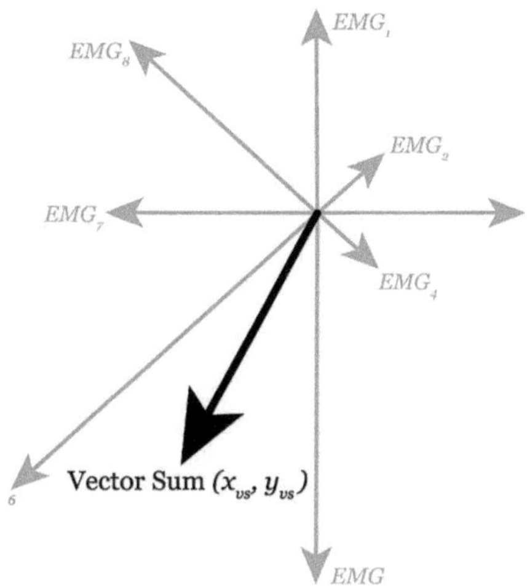

Fig. 5 A sample vector sum, drawn with the component EMG vectors (source: [47])

5 Making Sense of the Signal

After cleaning up the signal and extracting potentially useful features, the next step is to develop interaction, be it remotely controlling a device, changing visual output, or performing sound or interactive music. This can be done with the EMG alone, or in conjunction with other sensors in a multi-modal configuration. Motion sensors—IMUs made up of accelerometers, gyroscopes, or magnetometers—provide information that is complementary or supplemental with respect to the EMG from the same gesture, enriching the scope of interaction [25]. The activation of a particular muscle can be dependent on the relationship to gravity, as well as whether a pose is being held statically, or is in the context of a dynamic or "ballistic" gesture [26, 27]. Combining multiple channels of EMG with IMU sensors enables the interpretation of complex, expressive variations in gesture. This, however, creates an corresponding increase in complexity of information that needs to be aggregated and analysed. Automated forms of numerical analysis, notably machine learning, can provide powerful solutions in such situations.

Machine learning is a category of artificial intelligence that gives systems the ability to learn from examples rather than be explicitly programmed [28, 29]. A thorough understanding of the mathematics behind machine learning is not necessary to use machine learning in interactive applications. This section provides an overview of a workflow and algorithms that will allow readers to start experimenting with machine learning quickly and build up an intuition for what is possible.

IML is a specific approach to machine learning where a human provides one or more examples to a system in the form of "When I do X, please give me Y". X is a set of input features: sensor values or features extracted from the raw data. Y could be output parameters to control an audio process, switch on a motor, or generate an image. With the provided example, a learning algorithm attempts to train a model that will give results similar to the example data when provided with similar input and sensible responses when provided with totally novel input. The model should have learned what the user is expecting [10, 30, 31]. In an IML workflow, the human user then decides whether or not training has successfully created a model that is suitable for the application.

Contrary to the big data sets that are required for the mass effects of deep learning approaches, IML relies on relatively small amounts of data and algorithms that can train in a relatively short amount of time—seconds, rather than hours or days. Because models can be tested in action quickly, it is unnecessary to engage in the complex statistical validation of trained models that is typical in other types of machine learning. Instead, a user can rapidly train and test a model and then iteratively train again, with new data or even a different algorithm, until an acceptable result has been achieved.

Fiebrink and Caramiaux present a taxonomy of tasks suitable for artistic applications where IML can be used to *recognise* incoming actions, *map* input data to outputs, and *track* user activity and motion over time [32]. Here we apply this terminology to looking at EMG sensing of arm gesture in a multimodal context. We use tools that have been developed over the course of a series of research projects.[3,4] We also use techniques for gesture tracking developed at IRCAM [33]. These tools and libraries provide IML functionality for interactive projects and are available to the reader to use. They employ an IML workflow that allows users to quickly prototype new interactions and evaluate their performance through hands-on experimentation. The resources section, Sect. 6 below, covers practical details of these tools. In this section, we provide examples of how such tools might be employed and situations where IML can facilitate interactions.

5.1 Recognising

The most basic interaction with any sensor is to recognise that an action, or one particular action, has happened. In machine learning, this kind of evaluation would be called "classification", and a learned model tries to predict the correct label for a given input based on examples that had been provided during training. A recognised gesture can trigger an event.

5.1.1 Simple Recognition

With the EMG, recognition can be implemented by monitoring the amplitude (RMS or Bayesian filter) of a single EMG channel and checking to see if that signal

[3] http://www.rapidmixapi.com

[4] https://gitlab.doc.gold.ac.uk/rapid-mix/RapidLib

crosses a predefined threshold. When working with different people, each individual might produce a different range of EMG amplitudes. The most basic method of dealing with this variance is through calibration, allowing the user to set a threshold manually, choosing how much muscular activation will trigger an activation. A more automated method would be to measure the user's MVC and express the threshold as a percentage of the MVC value, for example, to trigger the desired event if the threshold passes 50% of the MVC. A generic interaction programme could be configured with a % of MVC, with personalisation taking place in an automated total MVC measurement at the beginning of the task as the sole calibration step. This can facilitate the case of complex interactions where there may be multiple trigger points tracking increasing muscle tension.

The EMG may be problematic in a threshold-based triggering application, given its noisiness. False or multiple triggers might result if the signal fluctuates around the threshold. A Schmitt trigger implements hysteresis in the triggering circuit as means for avoiding unintended triggering. The hysteresis loop sets different thresholds for triggering on and off, requiring the signal to descend below a lower threshold in order to "rearm" the upward-going trigger threshold. If, for example, the upward threshold is set to 50% of MVC, the system might be programmed to wait for the signal to go below lower threshold (perhaps 30%) before another upward trigger is enabled. This would help eliminate rapid oscillation between states. Rapid re-triggering is also a typical problem with mechanical switches and can be remedied by "debouncing", or waiting some fixed amount of time before allowing another event to be recognised. Depending on the project, either a time-based debounce or a dual threshold trigger might be called for.

In a situation involving multiple EMG sensors on a forearm, a project might want to distinguish between specific postures, like "closed fist", "open hand", or "pointing". Attempting to manually recognise them by setting thresholds for each individual EMG channel would be difficult to reproduce. In addition, these thresholds would need to be readjusted if the position of the electrodes changed or the sensors were placed on a different user. It is in such cases that an IML approach can be powerful, enabling a user to demonstrate the desired poses in a recorded training set that is analysed by an algorithm for subsequent recognition.

5.1.2 Classification with k-Nearest Neighbours

To recognise that something has happened in situations combining multiple sensors (either multiple EMG channels or EMG with other sensing modes) is a multidimensional problem. There could be different types (or classes) of events to be recognised. This is another case where manually hard-coding a recognition function would be extremely difficult and where IML could perform well. IML can be thought of as an "algorithm generator" which creates a model of the kind of signal processing that would require elaborate configuration if it were to be be hard-coded. Models can be generated on the fly, based on examples. Creating models with IML in cases like this will be quicker and more flexible than manual direct coding. The

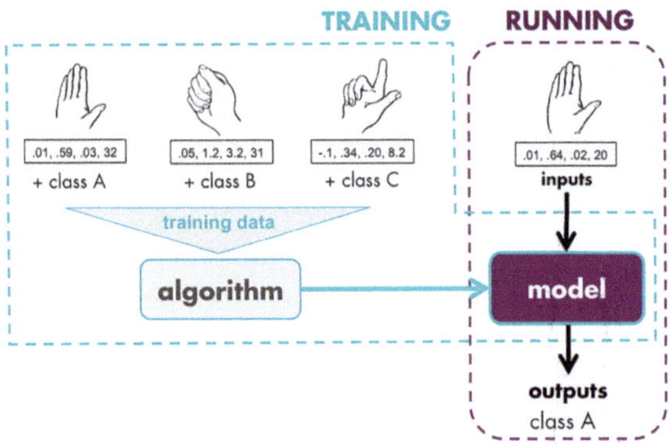

Fig. 6 An IML workflow: stages of training and testing or running a model. This model is recognising a class, as described in Sect. 5.1)

focus on creating examples and testing them keeps the user centred in the problem space of designing an interaction.

In machine learning terms, recognition is a "classification" task. One common classification algorithm is k-nearest neighbour, or k-NN. This algorithm compares an incoming signal to the examples it has learned (the training set) and returns the class associated with the most similar examples (see Fig. 6). With EMG and multiple electrodes on the forearm, typical classes might be different hand postures, like "fist" or "open hand". These classes could be represented by recordings of a range of sensor readings from the user performing these postures during the training phase. The k-NN algorithm creates the boundaries between recognising one posture and another. Generally, we start with $k = 1$, so the boundaries are defined by the single most similar example. The recognised result is reported as the class of the training example which has the smallest Euclidean distance to the new input.

The k-NN is relatively simple to setup and configure—k is its only parameter. There is no training time; therefore, it is very easy to add, replace, or remove examples. Finally, it works well in a large range of use cases. It is a type of supervised learning, where the labels indicating the class of each element in the training set must be given a priori.

Disadvantages of the k-NN include the fact that it is susceptible to noise (especially when $k = 1$). It can become computationally expensive if the number of examples and the number of feature dimensions are large. Also, it is an algorithm, when running, that is always classifying. That is to say, the model will return the best guess for a class, even if the guess is not very good. Unlike other algorithms, there is no default way of understanding the quality of the guess. So, the "nearest" class could be quite far from the test example.

There are several things a user can do to maximise the effectiveness of the k-NN with the EMG. Noise is characteristic of EMG signal and can pose a problem for k-NN. Noise reduction then becomes an important pre-processing stage when using this algorithm with EMG data. The envelope following mentioned above can be an important first step. Low-pass filters are the typical approach to noise reduction. There are three places to consider filtering. First, when collecting examples. It is

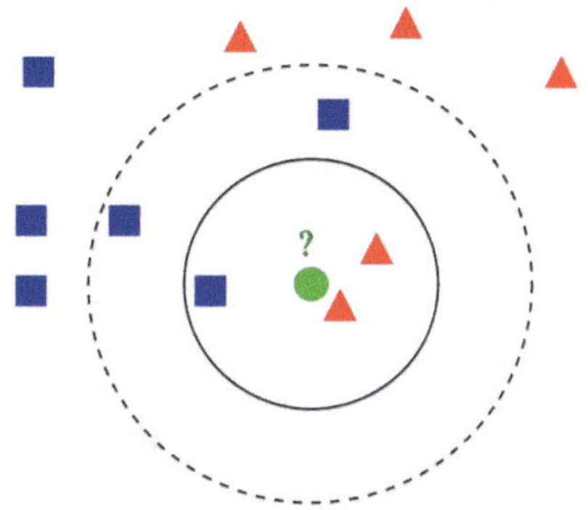

Fig. 7 Classifying with a k-NN model (Image source: https://commons.wikimedia.org/wiki/File:KnnClassification.svg)

typical to provide multiple examples of each class, especially with gestures that could be realised by diverse types of muscular activity. The ideal set of examples represents the full range of every class without any noisy data, but this is not always practical when monitoring streams of sensor data. A low-pass filter when collecting examples is a good idea to make sure only representative examples are associated with each class. Recorded examples can also be edited after the fact, but that is potentially tedious, undermining the advantage of an IML workflow. If an application must work with a large set of noisy examples, it is more efficient to raise the value of k, from one to three or possibly higher. As can be seen in Fig. 7, the green, circle example will be classified with the red triangles for k values of one or three, but with the blue squares for $k = 5$.

Another place for potential filtering is on the data stream that is tested against a trained k-NN model. Noisy input could lead to mis-classifications. In practice, though, we have had better results filtering the *output* of a k-NN model. This will remove any momentary mis-classification caused by either the input signal or the training set and usually leads to a more usable result.

Another potential source of ambiguity, or noise, comes from the design of gestures to classify. We discuss open hand vs fist because these gestures are associated with very different muscular activation profiles and will give dependable classification results. Physical movement that looks clear to our eye might be producing ambiguous sensor data. It is good practice to visualise the sensor signal for each class to be recognised and verify that the type of data for each has a strong individual profile. If not, then the desired interaction might need to be reconsidered.

We often design systems that should do nothing if no particular gesture is happening. For example: play a drum sample if I make a fist or play a string sample if I open my hand—otherwise, be silent. If we just record examples of a fist or open hand, then that will be all the model outputs and there will never be silence. In such cases it is necessary to record a "null" gesture. These are examples of when nothing is happening, such as a completely relaxed hand/forearm. Recording examples of every variety of non-gesture can be impractical. An alternative approach is to combine k-NN with some adjustable thresholding technique. A system could be set to classify only in the case where there is muscular activity beyond a certain level.

Or, a different algorithm with gives strengths of potential matches, or probabilities, could be used.

5.1.3 Probabilistic Classification with Mixture Models

Gaussian Mixture Models (GMMs) (Fig. 8) are a class of probabilistic models that can be used to classify input. The value of a parameter in a class is put at the centre of a Gaussian curve representing the likelihood that a test parameter of a particular value matches that class. The closer a test parameter is to an example, the more likely it is part of that class. GMMs return a value or values that correspond to how likely a given input matches one or more of the classes in the training set. These probabilities can be useful in determining whether or not a good match has occurred. If the probability of the best class is low, or the input has similar probabilities for multiple classes, then a further analysis step might conclude that no class has been sufficiently matched and trigger no response. This could eliminate the requirement of a null gesture class, something that we showed to be useful or even necessary with k-NN. There is also more possibility to adjust, or "tweak", GMMs [34], whereas k-NN models are defined wholly by the example set.

Its probabilistic basis means that GMM can offer a contrasting set of strengths and weaknesses to k-NN. Strengths of GMM include the fact that trained models are very efficient. The probability of matching more than one class can be useful depending on the application. Because each mixture model in a GMM algorithm can be statistically generated from multiple examples, the resulting classification is less susceptible to noise than k-NN.

Weaknesses of GMM include practical issues such as the fact that training a GMM algorithm can be computationally intensive and be demanding on slow processors. GMM is not as effective when there is a high number of input dimensions.

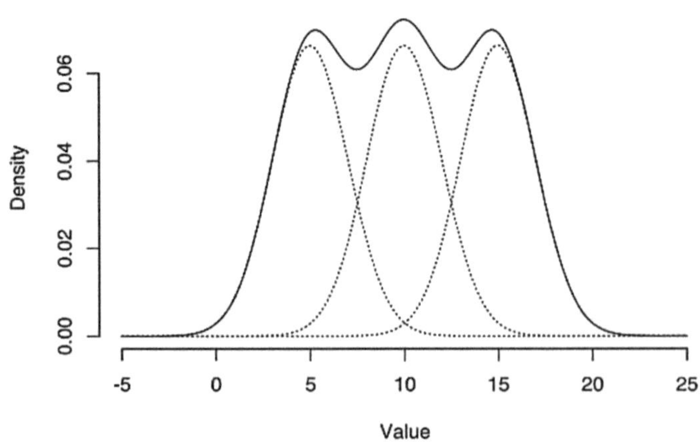

Fig. 8 Example of a Gaussian mixture model

There is a trade-off between computational cost in training and efficiency at run time. With a large number of examples, trained models can run more efficiently than a simple k-NN.

5.2 *Mapping*

Recognition is a discrete process: a fist is either open or closed, a "gesture" we may be classifying as one posture or another. Expressive musical systems depend on continuous interaction between input and output. One might use muscle tension to make a remote controlled vehicle go faster or slower, make a light brighter or dimmer, or shape a sound synthesis parameter, such as a filter, continuously. Continuous interaction creates a different problem space than interaction that follows a simple recognition and triggering paradigm. This kind of control can be described as "mapping" one or more input signals to one or more output parameters.

5.2.1 Manual Mapping

Miranda and Wanderley [35] introduced several canonical mapping strategies in their discussion of musical interaction. Strategies can be characterised by the number of input channels and their mapping to control one or more musical output parameters. A one-to-one mapping uses one input value mapped to one output. As interactions become more complex, one-to-many, many-to-one, and many-to-many approaches are also valuable approaches.

A simple, direct mapping could be implemented using muscle tension. The amplitude of a single EMG signal could be mapped to the position of a servo or stepper motor, allowing the user to control the position of a robotic arm based on the level of muscle tension. Rather than mapping 0–100% amplitude directly to 0–100% extension, it might be desirable to pass the amplitude through a transfer function. For example, if the arm seemed to extend too suddenly, it could be useful to map EMG amplitude through an exponential curve. If input and output are both normalised to a range from 0 to 1, then the equation $output = input^2$ would yield a mapping that starts slowly and then increases more rapidly at the top of the range.

This kind of direct mapping with a mathematical transfer function is relatively straightforward in a one-to-one case. However, in many-to-one and many-to-many situations, hardwiring connections and defining the logic and mathematics of those connections are not always obvious. In those cases, machine learning approaches can offer an effective alternative.

5.2.2 Regression with Neural Networks

Regression is the machine learning term that corresponds most closely to the mapping task [36]. The process of creating a model that does mapping between complex inputs and complex outputs is similar to the training/testing paradigm from classification. The user provides a set of examples: when the sensor signals are *this* (input), then the result should be *that* (output). The regression algorithm then constructs a model that should closely match the provided examples, but also generates sensible output when the input signal is different than any of the examples. The model is, in effect, a mapping that predicts the entire relationship between inputs and outputs based on the set of examples without having to explicitly programme every possibility.

In practice, one could take EMG amplitudes from two or more sensors placed on different arm muscles, combine that with IMU data and other derived features, and use these data to create a multimodal, multidimensional model of gestural input. That model might be used to control multiple parameters of a musical performance such as tempo, amplitude, sweeping filters, and other synthesiser parameters.

Multilayer Perceptrons (MLPs) (Fig. 9) are a specific type of artificial neural network that are useful in creating regression models. An MLP contains multiple layers of "nodes", each of which apply a weight and a transfer function to the signals it is connected to, passing the result to the next layer, and finally the output. To model a specific mapping, weights are refined during a training stage with examples. An MLP can map any number of inputs to any number of outputs. In the implementations we provide in the resources section, one MLP is used for each output so that unrelated outputs do not influence one another, as would be the case if they were in a shared network. The default architecture, having one input layer with one node per input value, one hidden layer that has the same number of nodes as the input layer, and a single-node output layer, is relatively simple but has lead to useful results in gesture/sound interaction. In a broad range of cases, this topology offers a good balance between training time and potential accuracy or expressivity in the final model. That basic MLP topology works in many cases without the need to adjust network parameters. Generated models are reasonably resistant to noisy

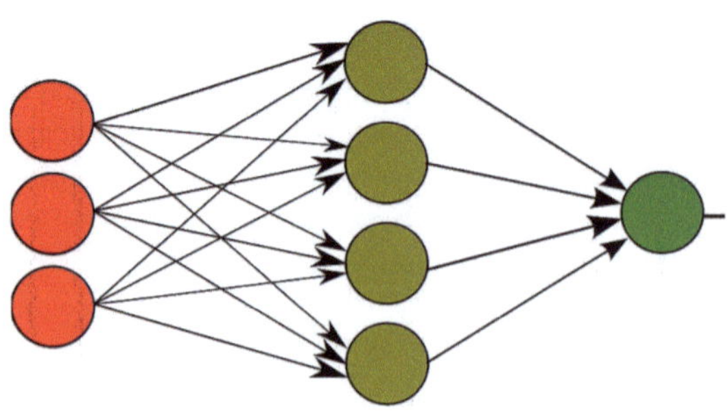

Fig. 9 A neural network configured as a basic multilayer perceptron (Image source: https://commons.wikimedia.org/wiki/File:MultiLayerPerceptron.svg)

training data, and though training can be CPU-intensive, a trained MLP model can be efficient enough to run on embedded or low-powered processors.

By automatically adjusting weights, an MLP will focus on the features that are correlated with changing outputs while ignoring features that are not. From the perspective of final model quality, it is usually better to include features that may or may not be useful and let the algorithm adjust appropriately. (Features do incur additional training time, however.) Finally, regression models can generate new outputs that are outside of the range of provided examples, thereby taking mapping outside an interpolation space, something we refer to as "extrapolation".

A weakness of MLP includes the fact that training time scales up rapidly with the product of the number of examples and the number of inputs and outputs. With small numbers of inputs (i.e. one or two), the basic MLP topology cannot simulate a very complex transfer function and may not provide smooth interpolation. If a trained model is not performing well, it is not always obvious if more training or different examples will help. Finally, there is no way to guarantee that a trained model provided by the neural network is the best possible model for the given input data.

By virtue of the algorithm, using an MLP for regression can generate a smooth transfer function. The training process is relatively resistant to noise. Unless there is a severe noise problem in the input signal, it is usually not necessary to filter out noise in the training set. Once a model is trained, however, noisy input will lead to noisy output. If that noise is a problem, undesirable noise can be filtered out either before or after the MLP in the signal path.

5.3 Regression with Gaussian Mixture Models

Although Gaussian mixture models were introduced as classifiers, it is possible to extend and configure them to perform regression tasks. This is called Gaussian Mixture Regression (GMR). With GMR, each class is associated with representative output parameters. The probability of each class (or the best n classes) is then used to weight each class' output parameters to create an interpolated output. GMR is very good with small data sets, and outputs will always be bounded by the parameters of the training classes. Disadvantages of GMR include the problem that model training time can be very long with many data points. The training time scales cubically with the number of data points. Also outputs will always be bounded by the parameters of the training classes, so extrapolation effects observed in our neural network regression implementation are not possible with GMR. We have applied GMR to study continuous variation in music conducting gestures between *legato* and *staccato* articulation [37].

5.4 Tracking

The systems for recognising and mapping described above do not handle temporal effects—they do not deal with actions that occur over time. Fiebrink et al. [32] introduces the notion of *tracking*, noting that "[s]ome systems that respond to human actions do more than just recognise that an action has occurred or map from a human state on to a machine behaviour".

Even though algorithms such as *k*-NN and MLP do not have any built-in notion of time or order, it is possible to incorporate information about temporal change into the training data. Accelerometer data from an IMU, for example, is data about how rapidly velocity is changing. This makes it very valuable for the gestural and performative interactions typical in the musical scenarios we are supporting. A static hand posture and the same posture as part of a dynamic gesture will have very different accelerometer data. First-order difference is another type of data where history can be built into the training data. Knowing that an EMG signal went from a very low voltage in one frame to a high voltage in the next could indicate that a percussive or plucking motion has just taken place and could be a useful trigger. Models can be trained on a window of samples to create a model that accounts for actions leading up to the current pose. The window size can be varied, for example, from 10 to 20 samples across a number of EMG channels, to isolate and detect the salient signature of a transient gesture as it is performed.

Temporal modelling algorithms are specifically designed to deal with time-series data and are useful for tracking tasks. They include Dynamic Time Warping (DTW) [38] and extensions of Markov chains including Hierarchical Hidden Markov Models (HHMM) [39]. They work by looking at a temporal gesture as a series of individual frames, or classification states. Incoming sensor and feature values can be tracked to distinguish between learned gestures.

Like *k*-NN, DTW compares an incoming series of states to the example series it has learned and returns the class of the most similar example. It performs recognition by looking for the lowest total distance between a sequence of data and each of the examples in a training set. This means that DTW is effective on sequences that may vary in speed (Fig. 10). A fast motion and a slow motion with similar muscle activity would be a strong match so long as their overall data trajectories are of a similar shape. This can be either a strength or a weakness, depending on the context of the gestures in an interaction. DTW classification is

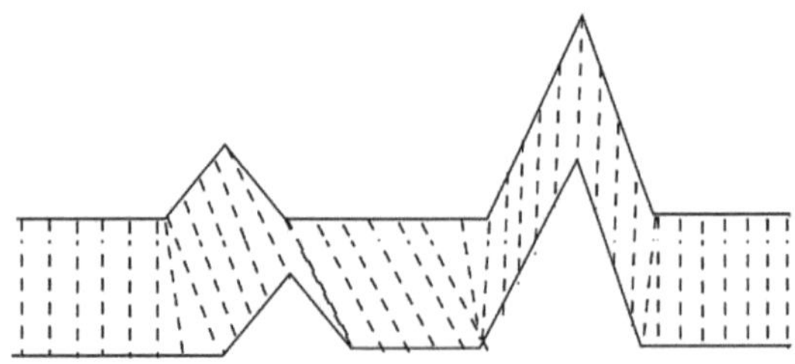

Fig. 10 Two sequences matched with dynamic time warping (Image source: https://commons.wikimedia.org/wiki/File:Dynamic_time_warping.png)

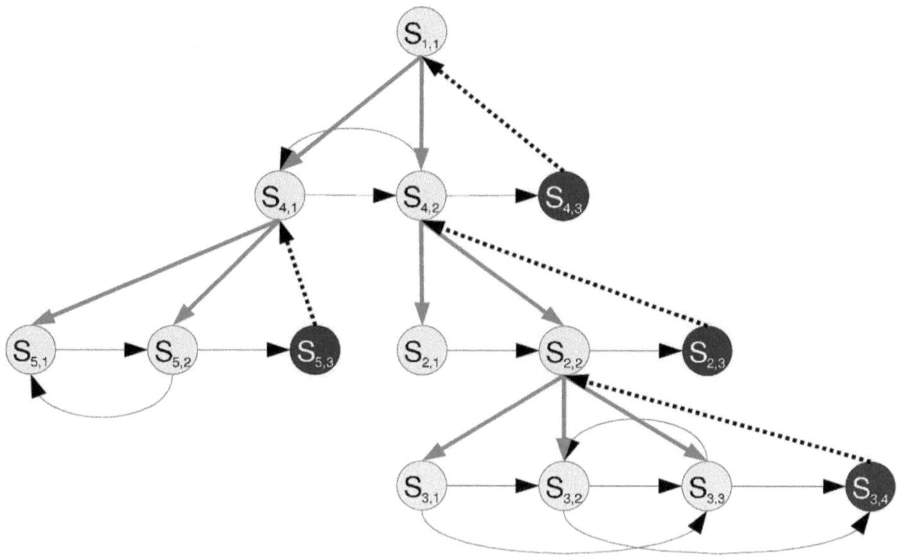

Fig. 11 States in a hierarchical hidden Markov model (Image source: https://en.wikipedia.org/wiki/Hierarchical_hidden_Markov_model)

strongest when it compares whole gestures, so input segmentation–identifying start and stop times—is important. Also, like k-NN, the closest match might not be a very good match if the input is very different from any of the training examples.

HHMM is an extension of temporal modelling by classical hidden Markov models to add a layer of classification for multiple gestures. While a classical HMM tracks the point in a sequence based on probabilities compared to a training example, HHMM also predicts what class a given gesture is in and reports what point in the current gesture is currently represented. If possible gestures were drawing shapes in space, such as a circle or square, then an HHMM model might give a result that indicates a user is 60% of the way through drawing a circle. This is especially useful for musical interactions where moving through a gesture could cause sonic output to progress. Often, as in the case with musical conducting, it is important to identify a gesture in progress because the next action will be triggered at the exact ending of the gesture. HHMM segments recordings of gestures into a series of parts, or states (see Fig. 11). The trained model calculates the probability of the current data representing a specific state in a specific gesture. This kind of probabilistic modelling has been applied to many problems in the musical domain [40].

6 Resources

The methods for building creative projects, described above, can be implemented using free, open-source software tools and resources. The authors are maintaining a supporting web page at https://mzed.github.io/RapidLib/IML-EMG_Strategies with a growing collection of links to libraries, examples, and documentation.

6.1 Data Acquisition

EMG and IML can be integrated into musical performance and interactive art using creative software environments, such as openFrameworks,[5] browser and server-based JavaScript, Unity, and Cycling '74's Max. This entails using communications protocols as pathways from hardware devices into each of these environments. Musical Instrument Digital Interface (MIDI) includes a standard serial communication protocol which enables communication between musical instruments and composition systems. Physiological data can be converted to formats prescribed by MIDI's event-based message syntax. Open Sound Control (OSC) [41] has been widely adopted in the field and supported in a large number of applications and BITalino R-IoT sends OSC natively over Wi-Fi.

The BITalino provides a C + + Application Programming Interface (API) which can be accessed from a creative coding platform using the same target language, such as openFrameworks. A programme written in openFrameworks can then be used as a bridge to send OSC to other applications, as demonstrated by software examples found on the resources page. There is also a publicly available add-on for using Myos in oF.

The EAVI BioMuse board [18], recently developed at Goldsmiths, sends MIDI over Bluetooth 4. EMG and IMU data are formatted to look like high-resolution (14-bit) pitch bend messages in MIDI. The device firmware is class compliant, making standard operating system service discovery to detect the EAVI BioMuse as a musical instrument input device. This eliminates the need for custom device drivers. MIDI is easily accessible in many creative applications, or could be converted to another format with free applications such as OSCulator.

Because of security concerns, getting data into a browser (and browser-side JavaScript) requires a web socket connection. This can be done with Node.js by redirecting incoming OSC messages to a web socket server. The ServerBIT application, covered elsewhere in this volume, will also forward data to browser code.

[5] https://openframeworks.cc/

Cycling '74's Max[6] is an environment for prototyping and developing data and media interactions. The resources page lists bespoke Max objects for using BITalino or Myo data directly in Max.

6.2 IML and Signal Processing

The interactive machine learning examples described above can be implementing using the RAPID-MIX tool set, either via individual components such as RapidLib and XMM or the main API available at http://rapidmixapi.com/. TensorFlow [42] was developed by Google as a means to implement big data and deep learning technologies. While the Python or C version is not oriented towards interactive workflows, Tensorflow.js[7] provides that possibility. The JavaScript-based ml5.js,[8] which builds on top of Tensorflow, makes those technologies available to creative coders accustomed to working in environments like Processing and p5js. Nick Gillian's Gesture Recognition Toolkit [43] provides an extensive C + + machine learning library and a GUI that supports interaction design.

Fiebrink's Wekinator software[9] is a free, open-source, standalone application (written in Java) that serves as an entry point into real-time, interactive machine learning. It uses OSC as inputs and outputs to enable multiple types of sensor and media interactions and provides high-level access to a number of established classification and regression algorithms.[10]

The RAPID-MIX consortium provides tools to implement IML on multiple platforms. Potential use cases include desktop and mobile platforms, embedded processors such as Raspberry PI, and web apps. Because the use cases are so diverse, libraries have been developed in both C + + and JavaScript. We started with Wekinator and XMM [33], which were used as points of departure in the development of a unified API. This API combines multiple interactive machine learning approaches and is designed to be friendly to creative DIY users [44]. As part of that API, we developed a lightweight signal processing and machine learning library called RapidLib, mentioned in the above examples.

RapidLib combines Wekinator-style IML with basic signal processing tools to implement sensor-based interactions in C + + and JavaScript environments. RapidLib provides *k*-NN, neural network, and dynamic time warping algorithms. While Wekinator is an effective tool for exploring IML, it is difficult to move from there to a production-ready application. RapidLib was born as a feature to export trained models from Wekinator as C + + code and is a one-to-one

[6] https://cycling74.com/products/max

[7] https://www.tensorflow.org/js

[8] https://ml5js.org/

[9] http://www.wekinator.org/

[10] https://www.kadenze.com/courses/machine-learning-for-musicians-and-artists-v/info

implementation of Wekinator's default classification, regression, and dynamic time warping algorithms. The current version of Wekinator will export models as JSON files that RapidLib can import.

To help with managing sensor inputs, RapidLib has a class called `rapidStream` which offers some basic signal processing such as RMS amplitude and low-pass filtering. Specifically for EMG, it implements the Bayesian amplitude filter described above. There is an openFrameworks add-on, ofxRapidLib, with examples for use in that environment.

The RapidLib library has been ported to JavaScript (using Emscripten [45]) to allow both browser and server-side interactions in JavaScript. There is documentation on the resources website and links to versions that can be included in a website or imported into Node.js. RapidLib has also been implemented as a Max object, RapidMax, and as a plugin for Unity, both of which can be found on the resources page.

XMM also implements classification and regression, providing statistical distribution and temporal modelling though implementations of Gaussian mixture models and hidden Markov models. It implements IML specifically geared towards motion recognition and mapping. XMM is also incorporated into the RAPID-MIX API, or it can be downloaded separately. In Javascript, it exists as an add-on package for Node. This allows models to be trained on a server and then be exported to run on a client, such as a mobile device.

In the signal processing domain, XMM pairs well with Pipo [46], which is a plugin API for processing streams of multidimensional data. This API was adapted by RAPID-MIX for implementing more advanced signal processing than rapidStream. Parts of XMM and Pipo are implemented in Max in the Mubu library. This includes XMM machine learning models and the Bayesian amplitude filter.

7 Summary

This chapter has demonstrated opportunities for creating accessible devices that harness sensing technologies and interactive machine learning for performing complex interactions in several domains. The work reported uses EMG combined with inertial sensing, which are a complex set of signals, to interact with music, which is a challenging domain. IML has made it possible for us design expressive interactions. These techniques were implemented primarily using free, open-source software and can be re-purposed by the reader for projects in their own chosen domain.

We introduced interactive machine learning as a workflow for designing complex interactions with combinations of sensors and especially EMG. As explored in Sect. 2, EMG is a highly nuanced and expressive biosignal. There are many methods for capturing this signal from one or more muscles on different parts of the body. We have focused our work on muscles in the forearm. However the signal is being captured, special care must be taken to avoid noise. This is a particular challenge for EMG. Even without external noise, the raw EMG signal is intrinsically quite noisy. Section 4 discusses approaches for extracting meaningful data from these signals. We have found estimating the amplitude envelope using a Bayesian filter to be effective in our work, allowing us to understand the state of a muscle quickly and accurately.

Use of IML to implement multiple interaction strategies for different expressive designs is the main focus of this chapter and is covered in Sect. 5. Strategies are considered in relation to specific interaction situations: recognition, mapping, and tracking. The appropriate strategy can be deployed based on the intended input gestures as well as the desired output (i.e. what is being controlled).

Recognition strategies identify when a gestural event (a fist, an open hand, a wave, etc.) has occurred and are supported by IML algorithms known as classifiers. For simple recognition, IML might not be necessary and more traditional signal processing will suffice. IML models offer advantages when dealing with multiple, complex inputs and can also provide details about the strength or likelihood of a specific prediction.

More continuous or nuanced interactions call for a *mapping* strategy. As with recognition, it is possible to implement mapping manually. However, with regression algorithms users can create complex designs with multiple inputs and outputs quickly and intuitively. Training by example, using IML, allows the designer to stay focused on the quality of the interaction, rather than the details of the code. Furthermore, regression models generate interaction spaces to explore, revealing new possibilities between and around the examples provided. Exploration can evoke unforeseen modes of interaction and prompt creative developers to new designs.

Tracking algorithms deal with gestures that take place over time. For these strategies, the designer might simply want to know which gesture (e.g. throwing a ball) happened. More expressive interactions can be developed if the model provides information about how the gesture was performed. Was the ball thrown quickly or slowly? What direction was the ball thrown? Models can predict that a gesture has, probably, started and report data about where in the gesture a user currently is. All of these data offer the potential for genuinely rich and natural interactions.

Tools for implementing machine learning have seen rapid evolution and refinement in recent years and currently offer a range of possibilities for creative use that would have previously been impossible. This field continues to grow, especially in the area of lightweight libraries aimed at non-expert users. The RAPID-MIX project was part of this expansion. Software packages from RAPID-MIX—RapidLib and XMM—are free and open source and can be implemented relatively easily in new applications, projects, or products. Links to these tools and examples are given in Sect. 6. The supporting website[11] will be maintained as our work progresses and new strategies are realised.

[11] https://mzed.github.io/RapidLib/IML-EMG_Strategies

Acknowledgments The research leading to these results has received funding from the European Research Council under the European Union's Horizon 2020 research and innovation programme (BioMusical Instruments, grant agreement no. 789825) and was supported in part by RAPID-MIX, an EU Horizon 2020 Innovation Action, H2020-ICT-2014-1, Project ID: 644862.

References

1. J. Wu, L. Sun, R. Jafari, A wearable system for recognizing american sign language in real-time using IMU and surface EMG sensors. IEEE J. Biomed. Health Inform. **20**(5), 1281–1290 (2016)
2. E. Costanza, S.A. Inverso, R. Allen, Toward subtle intimate interfaces for mobile devices using an EMG controller, in *Proceedings of the SIGCHI Conference on Human Factors in Computing Systems* (ACM, New York, 2005), pp. 481–489
3. E. Costanza, S.A. Inverso, R. Allen, P. Maes, Intimate interfaces in action: assessing the usability and subtlety of EMG-based motionless gestures, in *Proceedings of the SIGCHI Conference on Human Factors in Computing Systems* (ACM, New York, 2007), pp. 819–828
4. R. Nuwer, Armband adds a twitch to gesture control. New Scientist **217**, 2906 (2013)
5. A. Tanaka, B. Di Donato, M. Zbyszyński, G. Roks, Designing gestures for continuous sonic interaction, in *New Interfaces for Musical Expression* (NIME, 2019)
6. M. Zbyszyński, B.D. Donato, A. Tanaka, The effect of co-adaptive learning & feedback in interactive machine learning, in *ACM CHI: Human-Centered Machine Learning Perspectives Workshop* (ACM, New York, 2019)
7. M. Zbyszyński, B.D. Donato, A. Tanaka, Gesture-timbre space: multidimensional feature mapping using machine learning & concatenative synthesis, in *Computer Music Multidisciplinary Research* (2019)
8. A.F. Reimann, E.H. Daseler, B.J. Anson, L.E. Beaton, The palmaris longus muscle and tendon. a study of 1600 extremities. Anatom. Record **89**(4), 495–505 (1944)
9. P. Dourish, *Where the Action Is: The Foundations of Embodied Interaction* (MIT Press, Cambridge, 2004)
10. R. Fiebrink, Machine learning as meta-instrument: human-machine partnerships shaping expressive instrumental creation, in *Musical Instruments in the 21st Century* (Springer, Berlin, 2017), pp. 137–151
11. S. Katan, M. Grierson, R. Fiebrink, Using interactive machine learning to support interface development through workshops with disabled people, in *Proceedings of the 33rd Annual ACM Conference on Human Factors in Computing Systems* (ACM, New York, 2015), pp. 251–254
12. F. Bevilacqua, E.O. Boyer, J. Françoise, O. Houix, P. Susini, A. Roby-Brami, S. Hanneton, Sensori-motor learning with movement sonification: perspectives from recent interdisciplinary studies. Front. Neurosc. **10**, 385 (2016)
13. C.J. De Luca, The use of surface electromyography in biomechanics. J. Appl. Biomechan. **13**(2), 135–163 (1997)
14. A. Tanaka, M. Ortiz, Gestural musical performance with physiological sensors, focusing on the electromyogram, in *The Routledge Companion to Embodied Music Interaction* (Routledge, London, 2017), pp. 420–428
15. N. Ward, M. Ortiz, F. Bernardo, A. Tanaka, Designing and measuring gesture using Laban Movement analysis and electromyogram, in *Proceedings of the 2016 ACM International Joint Conference on Pervasive and Ubiquitous Computing: Adjunct* (ACM, New York, 2016), pp. 995–1000
16. H.J. Hermens, B. Freriks, C. Disselhorst-Klug, G. Rau, Development of recommendations for SEMG sensors and sensor placement procedures. J. Electromyogr. Kinesiol. **10**(5), 361–374 (2000)

17. J.V. Basmajian, C.J. De Luca, *Muscles Alive: Their Functions Revealed by Electromyography*, vol. 5 (Williams & Wilkins, Baltimore, 1985)
18. B. Di Donato, A. Tanaka, M. Zbyszyński, M. Klang, EAVI EMG board, in *Proceedings of the 2019 conference on New Interfaces for Musical Expression* (UFRGS, 2019)
19. M. Puckette, *The Theory and Technique of Electronic Music* (World Scientific Publishing, Singapore, 2007)
20. D. Hofmann, N. Jiang, I. Vujaklija, D. Farina, Bayesian filtering of surface EMG for accurate simultaneous and proportional prosthetic control. IEEE Trans. Neural Syst. Rehabil. Eng. **24**(12), 1333–1341 (2015)
21. T.D. Sanger, Bayesian filtering of myoelectric signals. J. Neurophysiol. **97**(2), 1839–1845 (2007)
22. A. Tanaka, Intention, effort, and restraint: the EMG in musical performance. Leonardo **48**(3), 298–299 (2015)
23. A. Phinyomark, R.N. Khushaba, E. Ibánez-Marcelo, A. Patania, E. Scheme, G. Petri, Navigating features: a topologically informed chart of electromyographic features space. J. Roy. Soc. Interface **14**(137), 20170734 (2017)
24. B. Caramiaux, M. Donnarumma, A. Tanaka, Understanding gesture expressivity through muscle sensing. ACM Trans. Comput.-Hum. Interact. **21**(6), 31 (2015)
25. A. Tanaka, R.B. Knapp, Multimodal interaction in music using the electromyogram and relative position sensing, in *Proceedings of the 2002 Conference on New Interfaces for Musical Expression* (National University of Singapore, 2002), pp. 1–6
26. B. Caramiaux, F. Bevilacqua, A. Tanaka, Beyond recognition: using gesture variation for continuous interaction, in *CHI'13 Extended Abstracts on Human Factors in Computing Systems* (ACM, New York, 2013), pp. 2109–2118
27. B. Caramiaux, N. Montecchio, A. Tanaka, F. Bevilacqua, Adaptive gesture recognition with variation estimation for interactive systems. ACM Trans. Interact. Intell. Syst. **4**(4), 18 (2015)
28. I.H. Witten, E. Frank, *Data Mining: Practical Machine Learning Tools and Techniques*, 2nd edn. (Morgan Kaufmann, San Francisco, 2005)
29. C.M. Bishop, *Pattern Recognition and Machine Learning* (Springer, Berlin, 2006)
30. S. Amershi, M. Cakmak, W.B. Knox, T. Kulesza, Power to the people: the role of humans in interactive machine learning. AI Mag. **35**(4), 105–120 (2014)
31. J.J. Dudley, P.O. Kristensson, A review of user interface design for interactive machine learning. ACM Trans. Interact. Intell. Syst. **8**(2), 8 (2018)
32. R. Fiebrink, B. Caramiaux, The machine learning algorithm as creative musical tool, in *The Oxford Handbook of Algorithmic Music*, ed. by (R.T. Dean A. McLean (Oxford University Press, Oxford, 2016), pp. 181–208
33. J. Françoise, *Motion-Sound Mapping by Demonstration*. Ph.D. Thesis, UPMC, 2015
34. H. Scurto, F. Bevilacqua, J. Françoise, Shaping and exploring interactive motion-sound mappings using online clustering techniques, in *Proceedings of the 2017 Conference on New Interfaces for Musical Expression* (NIME, 2017)
35. E.R. Miranda, M.M. Wanderley, *New Digital Musical Instruments: Control and Interaction Beyond the Keyboard*, vol. 21 (AR Editions, 2006)
36. I. Goodfellow, Y. Bengio, A. Courville, *Deep Learning* (MIT Press, Cambridge, 2016). http://www.deeplearningbook.org
37. A. Sarasua, B. Caramiaux, A. Tanaka, Machine learning of personal gesture variation in music conducting, in *Proceedings of the 2016 CHI Conference on Human Factors in Computing Systems* (ACM, New York, 2016), pp. 3428–3432
38. S. Salvador, P. Chan, Toward accurate dynamic time warping in linear time and space. Intell. Data Anal. **11**(5), 561–580 (2007)
39. J. Françoise, A. Roby-Brami, N. Riboud, F. Bevilacqua, Movement sequence analysis using hidden Markov models: a case study in Tai Chi performance, in *Proceedings of the 2nd International Workshop on Movement and Computing* (ACM, New York, 2015), pp. 29–36
40. D. Temperley, *Music and Probability* (MIT Press, Cambridge, 2007)

41. M. Wright, Open sound control: an enabling technology for musical networking. Organ. Sound **10**(3), 193–200 (2005)
42. M. Abadi, P. Barham, J. Chen, Z. Chen, A. Davis, J. Dean, M. Devin, S. Ghemawat, G. Irving, M. Isard, et al., Tensorflow: a system for large-scale machine learning, in *12th USENIX Symposium on Operating Systems Design and Implementation (OSDI 16)* (2016), pp. 265–283
43. N. Gillian, J.A. Paradiso, The gesture recognition toolkit. J. Mach. Learn. Res. **15**(1), 3483–3487 (2014)
44. F. Bernardo, M. Zbyszyński, M. Grierson, R. Fiebrink, Evaluating the usability of an API for rapid prototyping music technology with interactive machine learning. Front. Artif. Intell. **3**, 13 (2020)
45. A. Zakai, Emscripten: an LLVM-to-Javascript compiler, in *Proceedings of the ACM International Conference Companion on Object Oriented Programming Systems Languages and Applications Companion* (ACM, New York, 2011), pp. 301–312
46. N. Schnell, D. Schwarz, J. Larralde, R. Borghesi, Pipo, a plugin interface for afferent data stream processing operators, in *ISMIR* (2017), pp. 361–367
47. A. Tanaka, F. Visi, B. Di Donato, M. Klang, M. Zbyszyński, An end-to-end musical instrument system that translates electromyogram biosignals to synthesized sound. Computer Music J. **47**, 1–40 (2024)

REPOVIZZ: An Online Platform for Open Access and Visualization of Time-Aligned Multimodal Data Streams

Esteban Maestre and Oscar Mayor

1 Introduction

1.1 The Need for Open, Online Access to Multimodal Biosignal Data

Today's increasing affordability of technologies for biosignal data recording should be matched by improvements in documented archiving and in the remote visualization, exchange and retrieval of collections of recorded data. To exploit the wealth of available data in both commercial and academic contexts, the community must first devise frameworks that support the flexible exchange and enhanced browsing of biosignals in multimodal data collections. Furthermore, we should do so in such a way that showcases applications and makes them reproducible. Such an endeavor, which involves acquiring and describing recorded data, as well as remotely accessing and visualizing the data, needs daring, early-stage pursuits that inspire members of the community to design new tools for interactive, web-based consumption of multimodal biosignal data. These tools and initiatives should benefit not only future collaborative research but also commercial or entertainment-focused applications. Toward that end, in the context of multimodal music performance recordings (audio, video, motion capture, motion descriptors, physiological signals, annotations, etc.), we developed a biosignal-compatible hosting platform through which multimodal data can be stored, visualized, annotated, and selectively retrieved

E. Maestre
McGill University, Montréal, QC, Canada

Universitat Pompeu Fabra, Barcelona, Spain

O. Mayor (✉)
Universitat Pompeu Fabra, Barcelona, Spain

H. P. da Silva et al. (eds.), *Open Source Biomedical Engineering*,
https://doi.org/10.1007/978-3-032-03655-1_6

via a web interface and a dedicated API: the Repovizz system (https://Repovizz.upf.edu) [1].

In the audio/music community, the idea of developing tools for remote access and exchange of content is not new. But as the number and nature of acquired signals, modalities, extracted descriptors, and annotations increase, difficulties arise not only in the design or adoption of formatting schemes that allow the interchange of complex multimodal data collections, but also in the availability of platforms that enable and facilitate such an exchange by providing simple remote means to visualize and retrieve data. A simple, flexible model for formatting and organizing multimodal datastreams, together with an online system through which it is possible to host, preview, and download large sets of multimodal data, is an attractive asset in research and entertainment contexts. In addition to available entertainment-focused web applications such as YouTube or Soundcloud, it is possible to find research-oriented platforms, such as Free-sound, The Echo Nest, or AcousticBrainz, which provide an extended means for computing audio descriptors, conducting tag-based searches, and previewing audio signals. However, such platforms offer limited flexibility for remotely previewing extracted descriptors or annotations.

Beyond audio, it is difficult to find online data-exchange frameworks that offer rich interaction mechanisms. For example, the ENST-Drums dataset [2] can be accessed, a rich collection of annotated audiovisual recordings of drum performances for which the remote access mechanism is limited to downloading video, audio, and text files. The limited availability of remote visualization mechanisms for mocap data or physiological signals highlights the problem. You can find several web-accessible mocap databases of everyday human gestures [3], or even violin bow strokes [4], for which interaction is limited to download links, and no preview is available. So, you often need to download the data to a local drive and open it with specialized software to explore it or work with it. This issue, which evidently extends to datasets of recorded biosignals (such as Statistical Parametric Mapping datasets; www.fil.ion.ucl.ac.uk/spm/data), could be addressed if initiatives like Physionet [5] took full advantage of state-of-the-art web technology to offer unified, user-friendly means for browsing, visualizing, annotating, and contributing data.

2 Motivation

2.1 *What Led to the Creation of Repovizz?*

In Fig. 1, we have schematically illustrated a generic ecosystem for online open access and visualization of multimodal biosignal recordings. In this ecosystem, time-aligned biosignal recordings are preprocessed and analyzed to derive descriptor signals and annotations. Altogether, the derived set of signal streams and annotations corresponding to a multimodal biosignal is pushed to a remote repos-

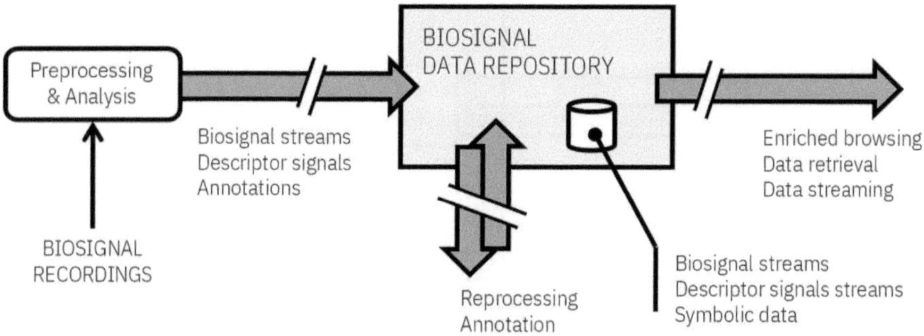

Fig. 1 An ecosystem for online access to multimodal biosignal recording collections

itory that holds indexed data streams made available for download, reprocessing, and re-upload. These data streams are also accessible through enriched browsing mechanisms that enable web previews of data as well as media or signal streaming as a means for open, effective navigation and interaction with data.

Departing from a music performance multimodal recording and analysis perspective, the Repovizz system reflects the efforts to realize one instance of such an ecosystem, considering music-related body motion and audio descriptors as generic biosignals. This research initiative began in 2006 at Universitat Pompeu Fabra in Barcelona, initially focusing on a novel paradigm for motion-driven synthesis of musical sound. This required creating a database of audio recordings, motion capture data, and sound and gesture descriptors that different researchers could annotate for internal use. Later, in the context of large-scale, international research projects focused on the study of ensemble music performance, the methodology for data acquisition and analysis was refined and extended to study a string quartet. Aligned with research needs, the vision for such a collaboration was to pursue the challenge of performing a series of large-scale multimodal recordings of quartet performances so that the recordings could be systematically processed, analyzed, and shared with the academic community through technical solutions that might inspire related applications with applicability in the biosignal domain. In doing so, a remote hosting platform and an enriched web interface were developed to ease open access and visualization, collaborative work, and reproducibility: the Repovizz system.

3 The Repovizz System

Instead of providing details about the software implementation of the Repovizz system, we briefly describe its structure and give an overview of the features it provides for potential uses and users. Repovizz lives in Barcelona and is hosted by Universitat Pompeu Fabra. It is composed of an API endpoint, a web server,

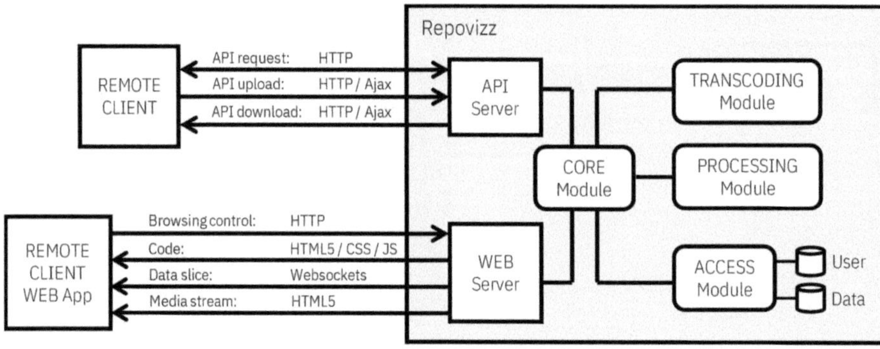

Fig. 2 Schematic representation of the Repovizz system architecture. It comprises an API endpoint, a web server, a transcoding module, a processing module, and an access control module, all connected to a central unit

a transcoding module, a processing module, and an access control module, all connected to a central unit. This is schematically illustrated in Fig. 2. Users can interact either programmatically through the API endpoint or through the web server using the web app.

Repovizz provides a cloud-based environment where multimodal, time-aligned data streams can be hosted and retrieved from a database of stream collections, or datapacks. An access control system ensures that datapacks marked as "closed" can only be accessed by collaborators, while datapacks published as "open" have no restrictions. Each datapack holds time-aligned streams (often corresponding to a single recording or to multiple recordings of the same duration), and the datapack is represented in the resource database as a tree document whose structure is freely defined by the data contributor.

The tree is formed by nodes that can organize data, hold textual descriptions, or hold pointers to data files of different types. Base types of nodes are generic signal nodes, audio nodes, video nodes, mocap nodes, and annotation nodes. Audio and video nodes hold pointers to audio and video streams uploaded in standard formats, while signal nodes are generic and can hold any type of uniformly sampled biosignal data stream, uploaded either in purposely defined ASCII or binary formats. Mocap nodes are a special type of signal nodes in which datastreams correspond to motion capture data consisting of 3-DOF tracked points and linking bones. Annotation nodes hold timestamp annotations in tabulated ASCII format.

Because multimodal data is often acquired in very-high-quality formats that allow detailed analysis but make it unfeasible to simultaneously browse or visualize data collections even if they are available locally, uploaded data is automatically transcoded or downsampled into lightweight formats and cut for a quick previsualization or streaming over the network. To provide agile means for remotely previewing and navigating dense data streams, we implemented data transfer on top of the WebSocket protocol. When previewing also implies playback of standard media streams, the Websockets protocol is combined with HTML5 media elements to enable synchronous waveform visualization and media playback. Conversely, the Ajax protocol is used when programmatic API commands request downloading a specific segment of one or various data streams in the originally uploaded format. Taking advantage of the imposed time alignment of all streams in each datapack, all requested datastreams can be automatically cut to within a selected time region and downloaded as a compressed Zip file containing both the tree document and the data files.

The Repovizz browser-based data navigation and preview engine lets users arrange and interact with various data visualization and media streaming widgets (such as multitrack audio waveform visualization, video playback, and 3D motion capture rendering and playback), purposely developed as HTML5 snippets delivered by the web server. Users can visually inspect data streams, synchronously play back media files, and edit and add time-stamped annotations in annotation nodes. Because remote collaborative work and reproducible research were two priorities in our vision, Repovizz provides data sharing features. For example, it can generate URLs with embedded access tokens that give access to protected resources, and it can link to specific web preview settings for a given datastream or dataset. In addition, a dedicated portal features a search engine that indexes data via diverse text identifiers, such as user-supplied tags, which aids data discovery. To empower Repovizz as a backend to support data access and visualization as companions to research articles or collaborative applications, we allow programmatic access to the system through its API endpoint. Using the HTTP-based REST API [6], clients can search for resources, fetch metadata, retrieve individual datastreams, and download datasets for further inspection. Repovizz also enables server-side data processing: for some modalities of data, a processing module running feature-extraction routines using the Essentia Library [7] can automatically compute and index descriptors upon upload.

4 An Example Use of Repovizz: The Quartet Dataset

Without loss of universality regarding its applicability to generic biosignals, we briefly describe an exemplary use case of Repovizz within our own research context: the Quartet Dataset [8]. The Quartet Dataset is a multimodal music performance dataset including multiple signals of different modalities, descriptors, and annotations whose compilation was motivated by a lack of ensemble music performance research based on analysis of time-aligned multimodal data signals recorded during music making. The Quartet Dataset focuses on the string quartet ensemble—a group of four musicians (two violins, a viola, and a cello) common in Western classical music practice. The dataset was originally created to study interdependence phenomena taking place between the members of a string quartet, through applying machine learning techniques in the analysis of recorded multimodal signals. We based the recordings on the following material: 9 exercises with a clear shared musical goal (for example, a common crescendo for all musicians) and 10 excerpts of musical pieces, where the shared musical goal is not stated but rather shaped throughout the rehearsals of the ensemble. Each piece or exercise was performed by the entire string quartet as well as individually by each musician. The performers were members of a professional string quartet. In total, the dataset comprises 96 synchronized, multimodal recordings that include:

- Multichannel audio data captured using ambient microphones as well as contact microphones placed in the bridge of the instruments
- Mocap data using two different high-end systems (one wired and one wireless), leading to 140 marker positions, including upper body motion and instrument tracking
- A video stream using a high-quality video camera

In addition to the acquired data, the dataset includes audio descriptor features as well as instrumental (sound-producing) gesture descriptors extracted from the mocap data and score-performance alignment data in the form of note onset and offset times. Within Repovizz, each multimodal recording in the Quartet dataset is stored as a single datapack. Besides the raw streams corresponding to audio, video, and motion capture modalities, each datapack contains computed audio descriptors, extracted segmentations as annotations, rigid body motion descriptors, and a set of derived bowing descriptors for each performer. For each exercise and piece or excerpt recording, five datapacks are provided: Solo violin 1, Solo violin 2, Solo viola, Solo cello, and Ensemble. Each datapack is identified by a short-form title. Figure 3 shows the web client interface to one of the datapacks, showcasing the data visualization and exploration capabilities of Repovizz for one of the experimental recordings. The datastreams in each datapack are organized into six grouping nodes:

- Recorded audio from each of the utilized microphones (cardioid and binaural ambient microphones, and a pickup microphone mounted on the instrument bridge for each performer), inside an AudioGroup node. Each audio recording is accompanied by the audio features extracted using the Essentia library in an adjacent AudioDesc node.
- Video stream, inside a VideoGroup node.
- Note onset and offset times for each instrument, obtained by score-performance alignment, in a Score node.
- Bowing gesture features for each instrument computed from motion capture data, in a DescriptorGroup node.
- Raw 6-DOF datastreams, acquired using the Polhemus motion capture system, in a MocapGroup node.
- Clean 3-DOF datastreams and rigid body metadata, acquired using the Qualisys motion capture system, in a MocapGroup node.

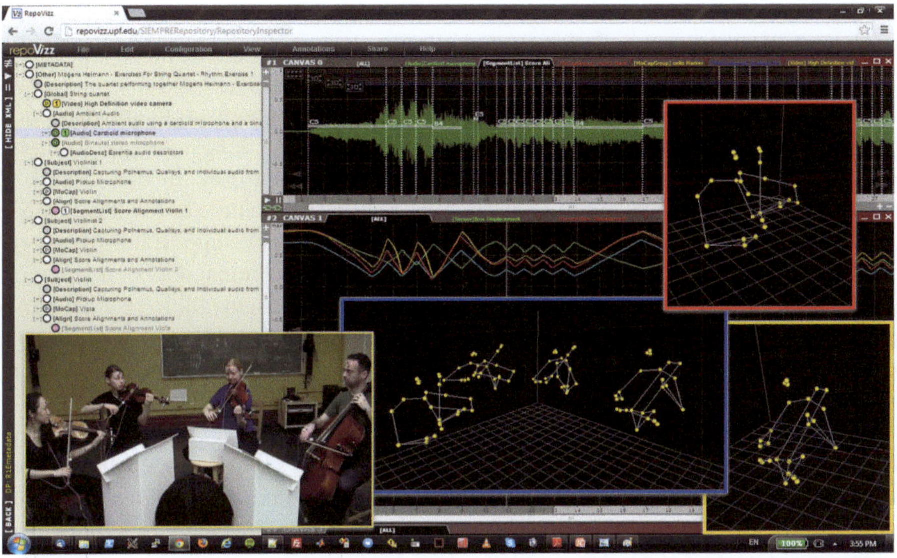

Fig. 3 A screenshot of the Repovizz web application interfaced to one of the Quartet Dataset recordings, showcasing the data visualization and exploration capabilities

All datapacks can be accessed online (http://mtg.upf.edu/download/datasets/quartet-dataset) or from the Repovizz portal, where a detailed tutorial of Repovizz helps users to get started in using the system.[1]

5 Reflections and Future Challenges

It is difficult to enumerate and justify here all the choices we have made throughout experiment planning, data acquisition, and platform development. Leaving aside classic challenges involved in data acquisition (such as synchronization), it was important to avoid uncommon, manufacturer-specific formats for data storage. A clear example is mocap, where manufacturers provide their own format and some tools to read and write their files. In our case, despite a loss in memory efficiency, we preferred to use plaintext CSV files as our basis for representation and storage, mainly to reduce eventual dependencies when sharing data. The same goes for other descriptor signals and annotations, which are also stored as CSV files with simple headers restricted to the sample rate when necessary.

[1] A tutorial with a step-by-step guide on using Repovizz is available online at https://Repovizz.upf.edu/misc/repoVizzTutorial.pdf

The choice of a tree document as the foundation for structuring time-aligned datastream collections was based on providing means for logical organization. An important aspect still needing a proper solution is the datatype definition. So far, we support a limited number of data types whose characteristics remain hardcoded within the system and used during retrieval and playback; thus, a service for user-defined datatypes would expand the flexibility of the platform. In that regard, if highly dense datastreams are to be used, a convenient alternative to the use of CSV files is currently under study (the HDF5 standard; https://support.hdfgroup.org/HDF5). In terms of collaborative work, we have yet to implement data or annotation versioning; when a user is accessing a datapack for visualization or interactive editing via the web app, its contents and editions are locked until the user saves his or her changes and leaves the datapack. To improve this, we are currently exploring the use of differential synchronization over a document-oriented database.

A yet-to-resolve issue is short-term scalability: at the moment, all data hosting and web serving is centralized and occurs at dedicated Universitat Pompeu Fabra webservers and computing infrastructure. If detaching and encapsulating the data serving processes from the main service, data server instances could be run within third-party online storage platforms (Amazon S3, for example), expanding possibilities for data decentralization and privacy enforcement. Finally, current web visualization and playback modes are limited to our own application needs and to the data types currently supported. In that regard, we plan to develop a widget development platform through which users can design visualizations and associate them with specific datatypes.

Although the developments presented here accommodated our own research needs and are allowing us to share our data, they only represent a generic attempt at expanding the possibilities for remote production-consumption of multimodal biosignal data. We hope Repovizz also serves as an inspiration for others in proposing and developing future means for exploiting the exchange of biosignal data in a digital library ecosystem that enables rich interaction with biosignal resources and fosters new forms of collaborative research or entertainment applications.

References

1. E. Maestre, P. Papiotis, M. Marchini, Q. Llimona, O. Mayor, P.A., Online access and visualization of enriched multimodal representations of music performance recordings: the quartet dataset and the repovizz system. IEEE Multimedia **24**(1), 24–34 (2017)
2. E.O. Gillet, G. Richard, ENST-drums: an extensive audio-visual database for drum signals processing, in *Proceedings 10th International Socity for Music Information Retrieval Conference (ISMIR* (2006), pp. 156–159
3. G.F. Torre, Guide to the Carnegie Mellon University multimodal activity (CMU-MMAC) database. Tech. report CMU-RI-TR-08-22,, Robotics Institute, Carnegie Mellon University, 2008
4. S.D. Young, A. Deshmane, Bowstroke database: a web-accessible archive of violin bowing data, in *Proceedings of the 7th International Conference* (2007), pp. 352–357

5. P.G. Moody, R. Mark, A. Goldberger, PhysioNet: a web-based resource for the study of physiologic signals. IEEE Eng. Med. Biol. Mag. **20**(3), 70–75 (2001)
6. Repovizz RESTful API documentation (Swagger UI). https://repovizz.upf.edu/repo/api-doc/index.html
7. D. Bogdanov, N, Wack, E. Gómez Gutiérrez, S. Gulati, P. Herrera Boyer, O. Mayor, G. Roma Trepat, J. Salamon, J.R. Zapata González, X. Serra, ESSENTIA: an audio analysis library for music information retrieval, in *International Society for Music Information Retrieval Conference (ISMIR 2013* (2013), pp. 493–498
8. MTG - Music Technology Group - QUARTET: multimodal string quartet performance dataset. https://www.upf.edu/web/mtg/quartet-dataset

Creating a Biomedical Dataset: Conceiving a Purpose Statement, Developing an Experimental Protocol, and Addressing Data Management

Ana Sofia Cacais do Carmo, Mariana Abreu, Joana F. Pinto, Tess den Uyl, Ana Fred, and Hugo Plácido da Silva

1 Introduction

Biomedical engineering is a broad field of practice, with a strong presence in both academia and industry, including areas of knowledge as diverse as biosensors, biosignal processing, biomechanics, medical imaging, and biomaterials [1]. As such, the need for experimental studies arises in the most varied forms. Although one is most likely very familiar with the concept of clinical trials—where an intervention is applied and the results are observed [2]—some studies (even preceding clinical trials) simply require the collection of real-world data involving human subjects. This is often seen in the domain of instrumentation, where sensor validation is a fundamental step, and/or machine learning, where the purpose is to identify (and then leverage) patterns from the real world (for which ground truth annotations are needed).

Nevertheless, just like in other research studies, the process of creating a dataset requires planning and follows the same principles:

1. Resultant data needs to be representative of the population under study, avoiding bias and enabling generalization of the outcomes.
2. Reproducibility is essential, not only from the perspective of the researchers, who should aim at performing every experimental session in the most similar

A. S. C. do Carmo (✉) · M. Abreu · A. Fred · H. P. da Silva
Instituto Superior Técnico, Universidade de Lisboa, Lisbon, Portugal

IT–Instituto de Telecomunicações, Lisbon, Portugal

J. F. Pinto
IT–Instituto de Telecomunicações, Lisbon, Portugal

T. den Uyl
University of Amsterdam, Amsterdam, Netherlands

H. P. da Silva et al. (eds.), *Open Source Biomedical Engineering*,
https://doi.org/10.1007/978-3-032-03655-1_7

conditions, but also so that the scientific community can validate the outcomes themselves.
3. Although not always possible, particularly in the corporate world, sharing the outcomes of the experiment not only increases the transparency and credibility of the findings, but also often contributes to scientific progress.

With these three requirements in mind, in this chapter, we seek to assist in properly designing and planning out a typical experimental study. In fact, this step goes beyond fulfilling a requirement when submitting a research study to an Institutional Review Board (IRB)[1] for approval, but instead provides a means to clarify one's ideas/purpose and to identify specific problems that may need attention [3].

Ranging from a student looking to collect data for an academic project, a researcher starting their experimental journey, or a startup planning to collect biomedical data, the goal of this chapter is to provide a set of guidelines and recommendations for the conduction of an experimental study that envisions the creation of a representative, reproducible and shareable dataset. Each section of this chapter addresses a crucial step, from its preparation and protocol design to the management and sharing of the resulting data, in a sequential manner. Moreover, each step is illustrated by the H2A2 Innolabs study—a collaboration between the IT (Instituto de Telecomunicações) Pattern and Image Analysis-Lx research group and the company VicarVision.[2]

The remainder of the chapter is organized as follows: Sect. 2 addresses the establishment of study requirements, Sect. 3 provides an overview of legal and ethical compliance, Sect. 4 discusses the actual design of an experimental protocol, and Sect. 5 addresses storage management.

2 Establishing Requirements

Establishing the requirements of the study is a crucial step, as it provides the starting point (and direction) for all other steps when conducting an experimental study, including the choice of an appropriate sample, participant recruitment, and design of the experimental protocol (addressed later in Sect. 4).

[1] An Institutional Review Board is an independent committee responsible for reviewing and issuing an assessment of the ethical soundness of a research study. In what concerns biomedical dataset creation, these committees exist to ensure protection of the rights and safety of subjects who participate in research studies.

[2] The H2A2 Innolabs study was setup by VicarVision, a computer vision company from the Netherlands, and was held in Instituto de Telecomunicações (IT), in Portugal.

2.1 *What Is the Purpose of the Study?*

When the need for data collection arises, one should start by defining the purpose of the study. While this chapter is more focused on the preparation and execution of an experimental protocol rather than structuring a research study plan, it is crucial to know the "why" behind that data collection. Whether the work aims at validating a product, comparing two methods, or assessing a research variable, without a clearly defined purpose as the foundation of the study, planning and implementing any of the remaining steps of an experimental study can easily deteriorate. For example, without the reassurance of a well-defined purpose, the researcher will most likely aim at collecting larger amounts and more types of data than required, out of fear of not collecting enough. While more data may seem tempting, this will only make both the data collection and data management processes less efficient.

While settling for a purpose statement or research question that will drive all remaining steps may seem like a daunting task, following the FINE criteria (initially proposed by Cummings et al. [4]) can assist in fulfilling this task. As such, a well-defined purpose statement should be:

Feasible

Important

Novel

Ethical

Moreover, the purpose of the study should be unambiguous and as specific as possible [5], specifying (whenever possible) the study design, outcome variables, and target population [3].

2.2 *Which Variables Will Be Studied?*

Another relevant matter to consider before attempting to develop an experimental protocol is the choice of variables to measure, which highly depends on the purpose of the study. For example, in exposure-outcome studies, both the factor to which a participant is exposed and the outcome are considered variables. The so-called confounding factors[3] should also be considered as variables to measure.

The concept of variable is not restricted to these types of studies, however. For example, when the purpose of a study is to characterize typical noise and artifact patterns in wearable electrocardiography during running activities, the

[3] A confounding factor is related to both the variable to which the participants are exposed and the supposed outcome.

study variables might be electrocardiography signals (of course), as well as other potentially informative factors such as electrodermal activity and accelerometer, which may be used to characterize that noise.

2.3 How to Enroll the Study Participants?

Before jumping into planning how to enroll colleagues and family members to participate in the study, it is crucial to establish the target population, i.e., the collection of individuals from whom the study aims to draw conclusions [6]. For every research study, the target population will be discriminated by a set of specific characteristics, such as demographics, clinical conditions, or practiced activities, for example [7]. This set of characteristics is called selection criteria, and it can be as broad as "subjects within the age range of 18 to 80" or as narrow as "healthy subjects in the first year of college, who practice competitive swimming."

In general, one should start by defining the desired characteristics of the population (known as inclusion criteria) and then specify the characteristics that should limit their participation in the study (known as exclusion criteria). These often correspond to characteristics that preclude the participant from giving informed consent (e.g., incapacitating psychological issues or not understanding the language in which the study will be conducted), that hinder the quality of the data, or that somehow interfere with the outcome (e.g., excluding participants who are under other medication when studying the effect of a specific treatment on a disease). Nevertheless, as a rule of thumb, to increase the representativeness of the study, there should be as few exclusion criteria as possible (which may imply converting exclusion into inclusion criteria, when possible) [7].

In reality, one will most likely not be able to approach the whole target population. Instead, only a portion of that target (called population sample) will be accessible, usually limited by geographical and temporal constraints [7].

Additionally, it is important to assess whether individuals from particularly vulnerable groups should be excluded. These may include children, pregnant women, prisoners, individuals with cognitive impairments, populations at risk of transitioning into one of these vulnerable categories, or those in hierarchical relationships that may compromise voluntary participation.

2.4 How to Guarantee That the Sample Is Representative of the Whole Population?

Representativeness of the sample is a crucial aspect, as it dictates whether the findings/conclusions of a study are generalizable to the whole target population or are, otherwise, invalidated [7]. Choosing an appropriate sampling approach may contribute to the maximization of generalization. Browner et al. [7] extensively describe sampling techniques and how to choose the one that best fits a study.

During sampling, particularly when a convenience sampling approach is taken,[4] potential systematic biases that may affect its validity should be identified [6]. Volunteer bias, a common error in research, may occur due to intrinsic factors (healthier and more educated people are generally more likely to volunteer [7]) or due to an inadequate recruitment technique (for example, when assessing digital literacy of a specific population, if a digital recruitment tool is used, volunteers will certainly be biased towards higher literacy levels than the generality of that population [8]). Moreover, people who are difficult to reach and those who refuse to participate often present different characteristics from those who volunteer, which may also affect the validity of the sample (known as nonresponse bias) [9]. Hence, when such biases are identified, extra effort should be devoted to ensure representativeness of these population subgroups.

Finally, having a sufficient number of participants also reduces the biases introduced by mere chance, increasing the probability of having a representative sample.

Achieving a sufficient number of participants implies, first and foremost, the definition of what is *sufficient*, which is dependent on the study design. For example, in binary test outcome studies, this value may be estimated according to the desired performance metric and confidence level [10]. Moreover, if the study aims at diagnosing a particular outcome that is not known prior to the analysis of the data, the prevalence of that outcome must also be considered. Tools such as ClinCalc.com[5] may be worth exploring. Finally, the possibility of corrupted data should be taken into account, whether from technical issues during the acquisition or from a lack of protocol adherence, for example. Browner et al. [11] provide additional statistical techniques that can guide you in that task.

2.5 How to Recruit Participants?

As mentioned before, inadequate recruitment may significantly hinder the validity of the sample. Moreover, unexpected problems during recruitment may result in an utter sense of despair from the investigator, while the initial recruitment enthusiasm fades and the participant number remains stagnant [9].

Therefore, the need for appropriate dissemination channels must be acknowledged. Best practices include propagation on social media, both through personal

[4] Convenience sampling consists in choosing participants merely due to ease of identification and enrollment. While this is a valid approach, one should be aware that this may result in a significantly biased sample and proceed accordingly [7].

[5] ClinCalc.com is available on https://clincalc.com/stats/samplesize.aspx (last accessed on September 2025).

and institutional pages. Another simple dissemination tool that works out well is distributing flyers about the study, stating its aim, target population, and contact information. Both from experience and as described in literature [7], word-of-mouth dissemination through previous participants remains one of the most efficient ways to reach more potential participants.

In case the sample has a very well-defined profile (personal, clinical, or socio-economic wise), it might be fruitful to identify communities or enterprises that gather people with that profile. If the population is targeted at healthy young volunteers, it may be advantageous to exploit the academic network to spread the study and recruit volunteers. Conversely, if the target population has a very well-defined clinical profile, then it might make sense to establish a collaboration with one or multiple clinical institutions specialized in that clinical condition.

Despite not being a mandatory procedure in biomedical research, rewards have been widely recognized as a positive means to potentiate both participants' recruitment and their motivation during the activities of the procedures. These can range from financial (e.g., gift cards) to gifts (e.g., consumer electronics devices or tokens of appreciation).

Finally, as good practice, one may consider regularly monitoring participant adherence while recruitment is ongoing, through the use of a log book of some sort (e.g., a spreadsheet with participant demographics and how they fit within the subgroups of the study design). This will not only inform about recruitment status but also help to keep track of the representativeness of all variables and subgroups.

2.6 The H2A2 Innolabs Case Study

Figure 1 summarizes the requirements established for the H2A2 Innolabs study, including study purpose, variables to measure, and population selection criteria.

Consider the following statements:

1. *Recording facial expressions using video.*
2. *Measuring facial expressions through mobile video recording.*
3. *Can facial expressions, measured through mobile video recording, be a predictor of emotional well-being?*
4. *Can facial expressions, measured through mobile video recording, be a predictor of emotional well-being according to age in participants with different baseline stress levels?*

The first purpose statement is very broad, hardly providing any information regarding the relevance of the study, the methodology, or the final objective. The following statements provide incrementally more information, namely, the study design (i.e., *through mobile video recording*), outcome variables (i.e., *facial expressions as a predictor of emotional well-being*), and target population (i.e., *according to age in participants with different baseline stress levels*). This example process of the H2A2 Innolabs study is an illustration of how to reach a final

STUDY REQUIREMENTS
H2A2 Innolabs Case Study

Study Purpose:

Can facial expressions, measured through mobile video recording, be a predictor of emotional wellbeing in participants with different baseline stress levels?

Study Variables:

- Age group ([25, 55] years vs [56, 75] years) and HADS level (HADS < 8 vs HADS ≥ 8) as independent variables
- Emotional well-being as dependent variables

Inclusion Criteria:

- Age 25-75 years
- (Semi-)independent living situation

Exclusion Criteria:

- Current diagnosis of mental disorder
- Current intake of medication for cardiac conditions

Sample Size & Study Design:

- 50 participants for each of the 4 groups
- 2 x 2 design

Fig. 1 Requirements established for the H2A2 Innolabs study. *HADS:* Hospital Anxiety and Depression Scale

purpose statement: starting with a broad statement and iteratively refining it into an unambiguous and specific study purpose. You may also notice that the third and fourth statements become research questions rather than mere purpose statements. While this is not always required or even feasible, stating the study purpose in an answerable format may help with subsequent steps [5].

To address age and baseline stress levels as potential confounding factors when predicting emotional well-being from facial expressions, a two-by-two factorial study design was proposed, as it allows the evaluation of the effects of two independent variables (each with two levels) on a single dependent variable. Thus, it provides a means to assess the effects of age (as given by two age groups: [25, 55] years vs [56, 75] years) and baseline stress level (measured using HADS:[6] (HADS < 8 vs HADS ≥ 8) on emotional well-being.

[6] The Hospital Anxiety and Depression Scale [12] measures anxiety and depression in a general medical population of patients and was used in the H2A2 Innolabs study to group participants with healthy and heightened baseline stress levels.

3 Ensuring Legal and Ethical Compliance

Current European Union (EU) regulations and guidelines for research are explicitly based on human rights, established through the European Charter of Fundamental Rights [13]. As such, while preparing any biomedical research study conducted with human subjects, a set of formalities and requirements must be addressed to make it compliant with the current legal and ethical policies.

3.1 Does the Research Need to be Approved by an IRB?

When preparing any kind of research project proposal involving human subjects (e.g., application for funding), approval by an IRB or equivalent entity is generally required. This is done to ensure in an independent manner that the participants' rights are safeguarded and that the research is conducted according to the best scientific practices in effect. It is common practice to use your own institution as the first source of ethics expertise and subsequent approval (the entity name can vary according to the institution, but an "Ethics Committee," "Ethical Review Board," or similar should exist).

Each IRB typically provides its own requirements/guidance regarding ethical compliance. However, the EU has made available an "ethics self-assessment" guide [14], which should be used to identify the ethical requirements of particular research cases . Considering the scope of this chapter, the focus will be on Sects. 2 and 4 of said guide—*"Humans"* and *"Personal data"*, respectively.

3.2 How to Obtain Informed Consent?

No intervention or experimental protocol can be performed without the explicit (and informed) consent of the study participants. What makes it an *informed* consent is the full disclosure, by the investigators, of what the participation entails, potential risks and benefits of the study, and alternatives to participation [15]. This is achieved by two fundamental documents:

- **Consent Form:** A brief document, containing the important information about the purpose and methodology of the study, to be presented to and signed by the participants at the beginning of the study. It thus consists of two parts, the information sheet and the consent form, and might include several criteria. An example template is made available by the World Health Organization (WHO).[7]

[7] Template made available by the WHO at https://www.who.int/ethics/review-committee/informed_consent/en/ (last accessed on July 2022)

– **Consent Withdrawal Form:** There are certain types of studies in which the participants are more likely to quit during the experimental period, due to pain, schedule constraints, or simply personal reasons. There are also studies that deal with delicate data, e.g., facial images or psychological behavior, in which even when the participants complete the data collection procedure, they might not feel comfortable having their data used. Therefore, it is required to make available a document to formally acknowledge and register the resignation, as well as inform the participants of its existence.

Since decision-making capacity is the first requirement for informed consent, one should be able to detect a lack thereof [16]—which may include, but is not limited to, children and adults with advanced dementia. In such cases, consent must be provided by the participant's legally authorized representative, and it must be ensured that they have sufficient information to enable them to provide this on behalf of the participant. Regardless, dissent should be respected [14].

3.3 How to Pursue Privacy by Design?

As two of the fundamental principles laid out in the European Charter of Fundamental Rights, the rights to privacy and protection of one's personal data must be guaranteed at all times (i.e., before, during, and after the conduct of the experimental protocol). While privacy can be a subjective concept, in general, participants desire to maintain control over personal information and communications. Therefore, if any personal information is subject to study, it is expected to be treated confidentially. The technical framework and security measures designed to ensure confidentiality constitute the **Data Protection Protocol**.

Some challenges arise when creating a data protection protocol, namely, how to identify what comprises personal data and how to pursue confidentiality within the constraints of your research purpose. In general, personal data comprises any information that may be used to identify the participant. However, the EU [14] distinguishes a special category of personal data (formerly known as *"sensitive data"*), which requires more stringent data-protection safeguards (e.g., political, religious or philosophical beliefs, genetic or biometric data for the purpose of uniquely identifying a person, or data concerning health).

> [**Personal data** consists in information about a person so that they become someone] *who can be identified, directly or indirectly, in particular by reference to an identifier such as a name, an identification number, location data, an online identifier or to one or more factors specific to the physical, physiological, genetic, mental, economic, cultural or social identity of that natural person* (Article 2(a) EU General Data Protection Regulation 2016/679 (GDPR)).

So, how should one devise a data protection protocol that avoids/limits data protection issues? Complete anonymization of all collected data, which implies encryption or removal of any personally identifiable data, exempts the study from any other data protection regulations. However, often there may be some information (which you might not think of as personal) that still allows you to identify someone. For example, if the sample under study is a group of students of a particular class that has only one male student, and gender is disclosed, this information is enough to identify that specific male participant (i.e., it is a distinctive characteristic in the sample under study). On the other hand, pseudonymization consists in separating data from its direct identifiers so that identification of a person is only possible with additional information (e.g., pseudonym or unique identifier).

As a rule of thumb, it is advisable to at least attempt the concept of **Privacy by Design**, which strives to incorporate privacy protection measures from the moment data is collected through technology. Alkhariji et al. [17] provide a state-of-the-art review of privacy by design principles and methods.

4 Designing the Experimental Protocol

Within any research study, an experimental protocol is a set of actionable guidelines or an operations manual that envisions the guidance of the staff through every single procedure and activity necessary to achieve the research goal.

An experimental protocol is especially critical in collaborative research projects [18], where the experimental sessions can be performed at different sites and by multiple staff members. A well-established (and pretested) protocol sets the ground for reproducibility and standardization.

4.1 How to Write an Experimental Protocol?

An experimental protocol is essentially a recipe, enumerating the ingredients needed and describing the required steps and conditions in which to practice them. Likewise, it should include the following key components:

- **Materials:** List of all the materials/equipment needed to perform the experimental session. Preferably, it should be in the form of bullet points, a checklist, or similar.
- **Methods:** Sequence of events. Clear description of the steps that should be performed by the staff and participant, accompanied by the expected duration and explicit conditions under which they should be performed.

The sequence of events in the protocol should start before the actual experimental session begins and prior to the arrival of the participant. This encompasses the opportunity to instruct the staff on the setup and careful preparation of the session. This may include gathering the required material, double-checking devices, proceeding with a particular cleaning routine (if applicable), or setting up the physical space. If the experimental sessions are performed by others, it is also important to describe more administrative aspects, such as how to address the participants once they arrive or how to handle the collected data.

Regarding structure, there is no conventional format for the methods of experimental protocols; in fact, these will vary according to personal preference and type of experiment, encompassing approaches such as tables, diagrams, or bullet points. Regardless, it is preferred to write the guidelines in an imperative form, as Renu and Hanson [19] explain in their guidelines on authoring technical instructions (e.g., "*Explain* the purpose of the study to the participant." instead of "The purpose of the study *should be explained* to the participant."). Clarity is key.

4.2 How to Minimize Potential Issues?

Brainstorming can prove to be an invaluable ally in identifying problems that may arise prior to or during an experimental session. Although not as informative as actively going through the actual set of guidelines and testing out the protocol, conceiving potential issues and pitfalls may prevent future unwanted errors in the data collection. Examples such as anticipating the possibility of forgetting to charge the data acquisition devices or not printing enough questionnaires will improve the clarity and completeness of the protocol, as well as make the team more attentive to such occurrences during the actual experimental sessions. In fact, this perfectly leads to a crucially important component of a complete experimental protocol: a set of contingency measures.

A thorough experimental protocol should explicitly describe how the staff must behave whenever one of the steps within the protocol goes against the expected. Did the team forget to brief the participant? Did one of the recording devices stop working? If so, should the step be repeated? Is it possible to proceed to the next step? Should the session be terminated? This should also include guidelines on how to proceed on a more administrative level (e.g., how to address the issue with the participant or what to do with the already-collected data). Each phase in the experimental protocol should be accompanied by a **fault recovery measure** so that the possibility of tampered/biased data and lack of standardization is minimized.

Regarding a more hands-on approach to problem-solving, pretests and pilot studies allow the evaluation of the feasibility and efficiency of the proposed experimental protocol [18], as well as the identification of certain **systematic** issues that may arise during the experiments. These should be done prior to IRB submission, as they might uncover fundamental issues with the methodologies within the protocol (e.g., unfeasible schedules, ambiguities within the task descriptions, or similar). However, it is important to bear in mind that since this phase is performed prior to any recruitment efforts, issues will most likely be related to protocol feasibility, compliance by the staff, and space/equipment flaws (particularly if the target population has very specific requirements, such as medical conditions). As such, only in the next phase will it be possible to detect **domain-specific** problems or lack of compliance by the participants.

4.3 How to Deal with Unexpected Issues During an Experiment?

A thorough preparation prior to the implementation of the protocol, as well as brainstorming sessions to identify potential problems, goes a long way to minimize the emergence of issues during the experiments. Nevertheless, some most likely could not have been anticipated. This does not mean the experiment's success should be left to wishful thinking. Timely identification and characterization of these issues will provide an opportunity to deal with them and perhaps even prevent recurrence.

Once the experiments have begun, it is advisable to have implemented a (preferably automated) preliminary analysis of the collected data. Depending on the type of data, this may look like alerts for out-of-range, illogical, or missing values (e.g., human-reported input fields), preliminary quality-checking algorithms (e.g., range-specific features), or simply data visualization (e.g., biosignal data that is expected to have a specific morphology). This preliminary quality assurance will allow the identification of potential **sporadic** issues that can not be controlled (e.g., resulting from human error or unpredictability of the variable under study) as well as **systematic** ones (e.g., inability of the target population to perform a task). However, make sure that what you have identified as inadequate data is not simply a consequence of an *ill-informed expectation* on your behalf. Otherwise, attempting to correct the collected data can significantly bias the results of the study.

In the case of sporadic/random issues, if the quality control is performed simultaneously with the experiment, it may be possible to promptly correct the faulty condition, or it will simply allow immediate discarding of faulty data, reducing the effort of subsequent data cleaning [18]. Again, it is important to additionally include a fault recovery measure for these cases in the protocol. Likewise, at this stage, identification of systematic issues may also imply some (hopefully) minor revisions to the protocol.

Nevertheless, if it is found that the experimental protocol (and consequently the validity of the study) would significantly benefit from a major revision after the study has begun, this possibility cannot be excluded. It is important to bear in mind that it is most probably an expensive option (e.g., if it implies repeating all previous sessions). Major revisions, such as adding study variables, altering the experiment conditions or modifying the eligibility criteria, require the approval of the IRB (associated with formal amendment of the study protocol), particularly if they imply any increase in risk for the participant or alterations to the consent form [18]. Besides, the data acquired prior to and after the revision should be subject to additional and separate analysis to infer an adequate interpretation of the impact of said revision on the study findings and conclusions.

4.4 *The H2A2 Innolabs Case Study*

The H2A2 Innolabs study required each participant to complete two in-person acquisition sessions in order to allow for test-retest reliability, as well as to assess individual variation. The two sessions were to be performed within two weeks, and each would imply a commitment of approximately 30–45 minutes from the participant, as detailed in the proposed protocol (Table 1).

When one considers the purpose of this study, establishing the required materials becomes rather intuitive. For example, to assess the reliability of mobile video recording, both a mobile device with a built-in camera and a high-frame-rate camera (for benchmarking purposes) are needed. Moreover, to collect the participants' baseline stress levels, relevant demographics, and perceived emotional state, self-report electronic questionnaires are a prevalent approach [20]. Finally, to confirm the emotional state in a less subjective manner, physiological parameters (such as electrodermal activity and heart rate variability) can be recorded simultaneously. As such, the complete list of materials becomes:

- Tablet with built-in camera
- High-frame-rate camera
- Electronic questionnaires
- Electrodermal activity acquisition device
- Two disposable electrodes
- Photoplethysmography acquisition device
- Laptop with acquisition software

Regarding experimental conditions, the same intuitiveness arises: to facilitate quality video recordings, great lighting conditions are particularly relevant, as well as sound isolation to avoid disturbances from external factors (for which only the participant and a staff member could be present in the room).

Overall, the design of the study consisted of asking participants to fill out information on a tablet and to do activities they would normally do, such as reading a news article or watching a video. During these activities, the facial expressions and physiological signals of the participants were collected. Table 1 further details the sequence of events and sessions that compose the proposed experimental protocol.

5 Managing Data

Managing the acquired data can be seen as a minor concern, passable to be dealt with at a later stage, once the experimental sessions are complete. However, failing to perceive the importance of a properly curated approach and thought-out structure may result in loss of data and time-consuming tasks of reorganization and file transfers. Nonetheless, planning in advance how to address this management may be a rather complex and daunting task, particularly if multiple types of data are

Table 1 Experimental protocol for the H2A2 Innolabs study, comprehending two acquisition sessions for each participant (**S1** and **S2**). (✓) indicates if the corresponding activity should be performed in both sessions or only in a particular one

S1	S2	Staff	Participant	Duration
✓	✓	Check presence of equipment and battery levels of sensors		5 min
✓	✓	Place tablet and camera according to the experimental setup mockup and turn both on		<1 min
✓	✓	Welcome participant		<1 min
✓		Explain the purpose of the study and how the informed consent works		10 min
✓			Read and sign informed consent	2 min
✓	✓	Verify informed consent (only for **S1**)	Fill out questionnaires 1 and 2 on the tablet	10 min
✓	✓	Connect sensors to the participant		2 min
✓	✓	Start recording on the camera and tablet		<1 min
✓	✓	Start sensors acquisition using laptop		<1 min
✓	✓	Check morphology of physiological signals	Perform activity A1 on the tablet while being monitored	5 min
			Fill out questionnaire 3 on the tablet	1 min
			Perform activity A2 on the tablet while being monitored	5 min
			Fill out questionnaire 4 on the tablet	1 min
✓	✓	Stop recording on the camera and tablet		<1 min
✓	✓	Stop sensors acquisition using laptop		<1 min
✓	✓	Remove sensors from participant and discard electrodes		1 min
✓		Confirm date of second session with the participant		<1 min
✓	✓	Thank participant and take them to the door		<1 min
✓	✓	Check validity of acquired data, charge sensors, and turn off all devices		10 min

involved and shareability is a concern. Instead, applying a bottom-up approach: What types of data will be acquired? Can they be stored in the same format? If not, are they compatible to be stored in the same place? Once those questions are settled, then start to think of dataset structure, security, and shareability.

5.1 How to Choose the File Formats?

An experimental protocol in biomedical engineering can entail a great variability of data types, often within the same dataset. For example, one may need to collect the participants' answers to questionnaires while, at the same time, monitoring a biosignal time series. These types of data require completely different storage formats, as well as approaches to processing.

Data tables provide the simplest form of storing collected data, in which each row usually corresponds to a single participant, with the columns containing the corresponding participant-specific attributes [21]. This data structure is often a suitable option to directly record data such as the participants' identification and personal information, as well as to store responses to surveys and online forms. In fact, well-established online survey tools (such as SurveyMonkey and Google Forms) allow for exporting the results to a spreadsheet format, avoiding the need for a manual transcription step. Nevertheless, spreadsheets mostly describe one-to-one relationships, making them rather inefficient when multiple study factors are at play and a more complex analysis is expected [22]. In such cases, you may find it valuable to instead consider relational database frameworks (e.g., MySQL or Oracle), as they facilitate queries and capturing of one-to-many relationships [21].

Moreover, with more complex data such as time series, structures like data tables can become rather restrictive. In such cases, each patient and experimental session often implies the storage of multiple channels, which may differ in resolution, sampling frequency, and physical units, invalidating the use of software tools like spreadsheets for visualization, preprocessing, or analysis. Besides, spreadsheets have limited capability when it comes to dealing with large volumes of data; as an example, Microsoft Excel supports a maximum of 2^{20} rows (i.e., less than 20 min of data acquired at 1000 Hz) [23].

5.2 How to Deal with the Particular Case of Biosignals?

Various aspects related to time series and biosignal storage are discussed in the literature, with concerns often hovering around the multiplicity of sampling frequencies, resolutions, and physical units, particularly when working in multimodal approaches (i.e., in which several biosignal modalities are acquired simultaneously) [24]. Memory concerns also play a big part in choosing the format in which time series are stored. This is especially relevant for large files, in which cases it may

Table 2 Comparison of file formats for biosignals according to a set of core functionalities and technical features. Adapted from [26, 27]. Label: *(–) no support, (+) support, (++) very good support, (+++) excellent support*

File format	F1	F2	F3	F4	F5
CSV	–	–	+	+	–
EBS	–	++	–	++	–
EDF+	–	+	–	+	+
GDF+	+	+	–	++	+
HDF5	+++	++	+++	+++	+++
MAT	+	–	++	+	++
MFER	–	–	–	+	–
WFDB	–	++	–	+	+

be preferable to resort to a format that allows for parsing the file in batches (e.g., line-by-line or using a buffer), instead of loading the complete content into memory.

Finally, it is important to consider that there is some flexibility when it comes to storage: if the format in which the data was initially collected does not suit the research needs (e.g., the format provided by the acquisition equipment), it may be of added value to include an additional step of reformatting [25]. Plus, it may also be relevant to aggregate multiple data sources, such as adding participant demographics and event annotations to a biosignal record, which facilitates subsequent data classification and meta-analysis of biosignal data.

Table 2 summarizes some file formats commonly used to store biosignals, along with how they behave according to the following set of core functionalities and technical features, adapted from Carreiras et al. [26]:

- **F1. Access performance:** Read and write speed, nonsequential access, data compression, or similar features.
- **F2. Event support:** Events and annotations relating to the time span of the acquired biosignals, which are particularly important for machine learning applications, where this information can be used by an automated algorithm to learn the meaning of specific signal patterns.
- **F3. Extensibility:** The ability to easily add more data to a file.
- **F4. Time series metadata:** Defined as data about data, including participant demographics, experimental conditions, recording equipment, and other relevant information pertaining to the corresponding time series.
- **F5. Multimodality support:** The capability to store various signal types in a single container structure, possibly accommodating different physical units, sampling rates, or resolutions.

Ultimately, when deciding on a file format, make sure it provides adequate support for the software or programming language being used to conduct your analysis.

5.3 How to Structure and Organize the Dataset?

Whether data is to be stored locally or on a remote server, setting up a logical system for the collected files will minimize duplication and accidental overwriting, while also facilitating access to data by the research team or other researchers. **Hierarchical folder structuring** is a commonly used approach, in which the data files populate hierarchically more specific folders, according to a particular organization strategy.[8] This approach might look something like having a top-most folder for the study at hand, with subfolders for each data collection session, participant, or experimental condition containing all corresponding recordings, as illustrated in Fig. 2. Together with an adequate naming strategy, hierarchical structuring provides an easy-to-navigate dataset organization, as well as great adaptability and extensibility.

Moreover, the file naming strategy should be concise and yet descriptive and, most importantly, consistent. According to the particular case of your experimental protocol, it may include the following elements [28], often separated by an underscore (i.e., "_"):

- Project name or acronym
- Date or date range of data acquisition
- Type of data
- Experimental conditions
- Version number
- Participant ID
- Responsible researcher name or initials

Figure 2 also highlights the existence of a *README* file, as it is good practice to provide documentation on the actual dataset (although the format in which this is provided is flexible). This documentation may include an explanation of the structure of the dataset, a description of the contents of each type of file format, as well as a clarification on how to access that content.

Furthermore, whenever some kind of processing is performed on the raw data, it is important to maintain an unchanged version, as well as adopting versioning of the files whenever different processing methodologies are attempted [28], as this may provide valuable insights for future research or documentation attempts [21]. Finally, loss of data should be minimized by ensuring regular backups of the complete dataset, including a local copy.

[8] "Data Management Guide - Organising your data," University of Cambridge (available at https://www.data.cam.ac.uk/organising-storing/file-organisation-naming-version-control, last accessed on September 2025).

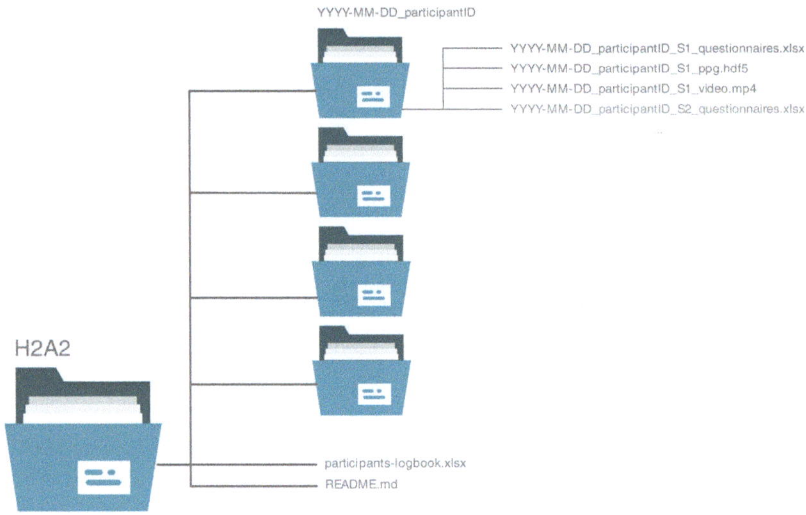

Fig. 2 Illustration of an exemplary hierarchical folder structure

5.4 How to Ensure Secure Shareability?

Sharing datasets is encouraged, as it not only increases the transparency and credibility of the findings by allowing other researchers to validate your methods, but also often contributes to scientific discovery way beyond its original purpose. Nevertheless, the shareability of the data should never be achieved at the expense of data confidentiality.

As discussed in Sect. 3, pseudonymization of the data goes a long way when it comes to pursuing confidentiality of the data, often providing sufficient means to meet data protection standards. That is, of course, assuming that the correspondence to these personal identifiers is either permanently eliminated after relevant information is extracted or at least maintained on a local storage or on a secure server, accessible only to authorized members [21]. Moreover, it is also important to decide if the entire dataset is to go public, if specific content should have restricted access (such as video footage or anything that may compromise the privacy of the participant), or if it is only intended to share it with other members of your research team. Either way, the metadata of the dataset should be made publicly available for transparency purposes.

When considering data sharing, the FAIR data principles should be considered [29]:

Findable

Accessible

Interoperable

Reusable

The FAIR Guiding Principles

To be Findable:

F1. (meta)data are assigned a globally unique and persistent identifier
F2. data are described with rich metadata (defined by R1 below)
F3. metadata clearly and explicitly include the identifier of the data it describes
F4. (meta)data are registered or indexed in a searchable resource

To be Accessible:

A1. (meta)data are retrievable by their identifier using a standardized communications protocol
A1.1 the protocol is open, free, and universally implementable
A1.2 the protocol allows for an authentication and authorization procedure, where necessary
A2. metadata are accessible, even when the data are no longer available

To be Interoperable:

I1. (meta)data use a formal, accessible, shared, and broadly applicable language for knowledge representation.
I2. (meta)data use vocabularies that follow FAIR principles
I3. (meta)data include qualified references to other (meta)data

To be Reusable:

R1. meta(data) are richly described with a plurality of accurate and relevant attributes
R1.1. (meta)data are released with a clear and accessible data usage license
R1.2. (meta)data are associated with detailed provenance
R1.3. (meta)data meet domain-relevant community standards

Fig. 3 The FAIR guiding principles, as proposed by Wilkinson et al. [29]

These are guidelines for the way data should be shared, with a particular emphasis on "enhancing the ability of machines to automatically find and use the data, in addition to supporting its reuse by individuals," thus benefiting researchers and aiming towards the maximum use and impact of your data. The FAIR principles are laid out in Fig. 3, but we refer the reader to the original publication [29] for a more comprehensive contextualization of the principles.

One way to share data is through an open research data repository. These will archive the data and often even allow assigning it a Digital Object Identifier (DOI) and a web page, displaying the information relevant for other researchers (i.e., metadata). The Teamscope website[9] describes several open data repositories and might help in finding a repository that will best suit the research team's needs. Moreover, there is a growing body of journals that are exclusively dedicated to describing and disseminating research data (such as **Scientific Data**, by Nature), exempting researchers from having to publish additional extensive research content

[9] https://www.teamscopeapp.com/blog/6-repositories-to-share-your-research-data (last accessed on September 2025)

around the dataset, thus facilitating data sharing. Finally, if the whole dataset is of sensitive nature, none of these solutions will be suitable, in which cases, one may instead offer to make the data available only upon request, binding the requester to the same consent signed by the participants.

6 Summary

In spite of the variety of paradigms in which the need for data collection arises, this chapter provides actionable guidelines in conceiving a purpose statement, developing an experimental protocol, and addressing the management of the acquired data. The following key points summarize the contents discussed throughout the chapter:

- The creation of a biomedical dataset should be built upon a **well-established study plan**, with a clear purpose or research question, target population, and appropriate recruitment approach.
- To ensure legal and ethical compliance, the study plan and experimental protocol need to be submitted to an IRB for approval.
- Current ethical regulations and guidelines advocate for confidentiality through **complete anonymization**, or at least **pseudonymization**, of the collected data.
- An experimental protocol should include a list of equipment needed, as well as a clear description of the steps that should be performed by the staff and participant (preferably in the **imperative form**).
- Managing the collected data includes exploring the most suitable file formats for each type of data. While spreadsheets may be a perfectly adequate option for patient demographics or questionnaire answers, other formats may be more convenient for time series.
- When choosing a file format to store time series, you may need to consider factors such as access performance, support for annotations and time series metadata, extensibility, and multimodality.
- Sharing datasets is encouraged, as it increases **transparency and credibility** of subsequent works. However, it should never be achieved at the expense of data confidentiality.
- The FAIR principles provide guidelines for sharing datasets, striving for **F**indable, **A**ccessible, **I**nteroperable, and **R**eusable data.

Acknowledgments This work was partially funded by FCT/MECI through national funds (FCT grant 2021.08297.BD and by FCT project reference and DOI identifier https://doi.org/10.54499/2022.12369.BD) and when applicable co-funded EU funds under project UID/50008: Instituto de Telecomunicações. It was also co-supported by the European Regional Development Fund (FEDER), through the Lisbon Regional Programme (LISBOA 2030) of the Portugal 2030 framework - LISBOA2030-FEDER-01318700 (ComSense).

References

1. J. Enderle, *Introduction to Biomedical Engineering* (Academic Press, New York, 2012)
2. W.S. Browner, T.B. Newman, M.J. Pletcher, Getting started: the anatomy and physiology of clinical research, in Hulley [S.B. Hulley (ed.), *Designing Clinical Research*, 5th edn. (Wolters Kluwer, 2022)], ch. 1 (2022), pp. 2–16
3. S.R. Cummings, A.M. Kanaya, Conceiving the research question and developing the study plan, in Hulley [S.B. Hulley (ed.), *Designing Clinical Research*, 5th edn. (Wolters Kluwer, 2022)], ch. 2 (2022), pp. 17–25
4. S.R. Cummings, W.S. Browner, S.B. Hulley, Conceiving the research question, in *Designing Clinical Research*, 1st ed edn. (Williams & Wilkins, 1988), pp. 12–17
5. T. Kwiatkowski, R. Silverman, Research fundamentals: II. Choosing and defining a research question. Acad. Emer. Med. **5**(11), 1114–1117 (1998)
6. D.A. Enarson, S.M. Kennedy, D.L. Miller, Choosing a research study design and selecting a population to study [Research Methods]. Int. J. Tuberculosis Lung Disease **8**, 1151–1156 (2004)
7. W.S. Browner, T.B. Newman, M.J. Pletcher, Choosing the study participants: specification, sampling, and recruitment, in Hulley [S.B. Hulley (ed.), *Designing Clinical Research*, 5th edn. (Wolters Kluwer, 2022)], ch. 3 (2022), pp. 26–38
8. X. Guo, E. Vittinghoff, J.E. Olgin, G.M. Marcus, M.J. Pletcher, Volunteer participation in the health eHeart study: a comparison with the US population. Sci. Rep. **7**, 1956 (2017)
9. M.X. Patel, V. Doku, L. Tennakoon, Challenges in recruitment of research participants. Adv. Psychiatric Treatment **9**, 229–238 (2003)
10. K. Hajian-Tilaki, Sample size estimation in diagnostic test studies of biomedical informatics. J. Biomed. Inform. **48**, 193–204 (2014)
11. W.S. Browner, T.B. Newman, M.J. Pletcher, Estimating sample size: applications and examples, in Hulley [S.B. Hulley (ed.), *Designing Clinical Research*, 5th edn. (Wolters Kluwer, 2022)], ch. 6 (2022), pp. 65–94
12. A.S. Zigmond, R.P. Snaith, The hospital anxiety and depression scale. Acta Psychiatrica Scandinavica **67**, 361–370 (1983)
13. European Union, Charter of Fundamental Rights of the European Union. Tech. Rep. C83/389–403, Official Journal of the European Union, 2010
14. European Union, EU Grants: How to complete your ethics self-assessment. Tech. Rep. V2.0, European Union, 2021
15. B. Lo, D.G. Grady, Addressing ethical issues, in Hulley [S.B. Hulley (ed.), *Designing Clinical Research*, 5th edn. (Wolters Kluwer, 2022)], ch. 7 (2022), pp. 95–114
16. B. Varkey, Principles of clinical ethics and their application to practice. Med. Principles Practice **30**(1), 17–28 (2021)
17. L. Alkhariji, N. Alhirabi, M.N. Alraja, M. Barhamgi, O. Rana, C. Perera, Synthesising privacy by design knowledge toward explainable internet of things application designing in healthcare. ACM Trans. Multimedia Comput. Commun. Appl. **17**, 62:1–62:29 (2021)
18. D.G. Grady, A.J. Huang, Study implementation and quality control, in Hulley [S.B. Hulley (ed.), *Designing Clinical Research*, 5th edn. (Wolters Kluwer, 2022)], ch. 18 (2022), pp. 323–339
19. R.S. Renu, L. Hanson, A rule-based decision support system for authoring technical instructions, in *ASME 2017 Int'l Design Engineering Technical Conferences and Computers and Information in Engineering Conference* (American Society of Mechanical Engineers Digital Collection, 2017)
20. C. Demetriou, B. Özer, C. Essau, Self-report questionnaires, in *The Encyclopedia of Clinical Psychology*, ed. by R.L. Cautin, S.O. Lilienfeld (John Wiley & Sons, London, 2014)
21. M.A. Kohn, T.B. Newman, Data management, in Hulley [S.B. Hulley (ed.), *Designing Clinical Research*, 5th edn. (Wolters Kluwer, 2022)], ch. 19 (2022), pp. 340–350

22. T.J. Dilling, Artificial intelligence research: the utility and design of a relational database system. Adv. Radiat. Oncol. **5**, 1280–1285 (2020)
23. D. Birch, D. Lyford-Smith, Y. Guo, The future of spreadsheets in the big data era, in *Proceedings EuSpRIG 2017 Conference on Spreadsheet Risk Management* (2018)
24. A. Schlögl, An overview on data formats for biomedical signals, in *World Congress on Medical Physics and Biomedical Engineering, September 7–12, 2009, Munich, Germany*, ed. by O. Dössel, W.C. Schlegel (Springer, Berlin, Heidelberg, 2010), pp. 1557–1560
25. R. Kraleva, V. Kralev, N. Sinyagina, P. Koprinkova-Hristova, N. Bocheva, Design and analysis of a relational database for behavioral experiments data processing. Int. J. Online Eng. **14**, 117 (2018)
26. C. Carreiras, H.P. Silva, A. Lourenço, A.L. Fred, StorageBIT: a metadata-aware, extensible, semantic, and hierarchical database for biosignals, in *Proceedings of the International Conference on Health Informatics (HEALTHINF-2013)*, ed. by J. Solé-Casals, A.L. Fred, H. Gamboa (SCITEPRESS, 2013), pp. 65–74
27. J. Rodrigues, H. Gamboa, V. Kublanov, A. Dolganov, Storage of biomedical signals: comparative review of formats and databases, in *2019 International Multi-Conference on Engineering, Computer and Information Sciences (SIBIRCON)* (2019), pp. 0652–0656
28. University of Glasgow, *Organising Your Data* (2020)
29. M.D. Wilkinson, M. Dumontier, I.J. Aalbersberg, G. Appleton, M. Axton, A. Baak, N. Blomberg, J.-W. Boiten, L.B. da Silva Santos, P.E. Bourne, J. Bouwman, A.J. Brookes, T. Clark, M. Crosas, I. Dillo, O. Dumon, S. Edmunds, C.T. Evelo, R. Finkers, A. Gonzalez-Beltran, A.J.G. Gray, P. Groth, C. Goble, J.S. Grethe, J. Heringa, P.A.C. 't Hoen, R. Hooft, T. Kuhn, R. Kok, J. Kok, S.J. Lusher, M.E. Martone, A. Mons, A.L. Packer, B. Persson, P. Rocca-Serra, M. Roos, R. van Schaik, S.-A. Sansone, E. Schultes, T. Sengstag, T. Slater, G. Strawn, M.A. Swertz, M. Thompson, J. van der Lei, E. van Mulligen, J. Velterop, A. Waagmeester, P. Wittenburg, K. Wolstencroft, J. Zhao, B. Mons, The FAIR Guiding Principles for scientific data management and stewardship. Sci. Data **3**, 160018 (2016)
30. S.B. Hulley (ed.), *Designing Clinical Research*, 5th edn. (Wolters Kluwer, 2022)

Using Open-Source Hardware for Enhancing Lean Entrepreneurship

Raffaele Conti and Salomé Azevedo

1 Introduction

The idea that entrepreneurship is plagued by uncertainty is not new. From Cantillon [1] to Say [2], continuing with Knight [3], Marshall [4], and Drucker [5], it has been widely acknowledged that the process of launching a new venture is essentially fraught with challenges and risks. Indeed, for entrepreneurs, it is typically quite difficult to know ex ante—i.e., before actually entering the market—whether their venture will be successful. This fundamental uncertainty affecting startup activity is confirmed by the highly skewed distribution of returns from entrepreneurship, with most ventures failing badly and only a small minority of them being very successful [6]. For instance, Shane [7], focusing on the about 500 000 new ventures created in 1996 in the United States, found that more than half of those ventures failed, whereas only 3500 achieved revenues greater than ten million dollars.

Both practitioners and scholars have tried to understand how entrepreneurs should cope with uncertainty, to reduce the risk of failure and increase the chance of success. In the broader management literature, several conceptual framings, such as the organizational learning perspective (e.g., March, 1991) or the real option theory (e.g., [8]), might in principle provide entrepreneurs with useful insights for navigating the uncertainty characterizing the process of launching a new venture—for a summary of these different theoretical perspectives, see [9]. However, these theoretical framings are usually quite abstract in nature, such that the insights they provide, even if valuable, might be difficult for practitioners to apply. Only recently,

R. Conti
ESSEC Business School, Cergy, France

S. Azevedo (✉)
Value for Health CoLAB, Avenida Fontes Pereira de Melo, Lisbon, Portugal
e-mail: salome.azevedo@vohcolab.org

H. P. da Silva et al. (eds.), *Open Source Biomedical Engineering*,
https://doi.org/10.1007/978-3-032-03655-1_8

the "lean startup approach" emerged as a methodology that is at the same time easy to teach to potential entrepreneurs and readily applicable by them, for coping with the uncertainty characterizing the entrepreneurial process [10].

The lean startup method points out that experimentation is key for solving the uncertainty inherent in entrepreneurship and that entrepreneurs should behave as if they were scientists. Hence, they should formulate hypotheses about how they intend to create value and test these hypotheses via experiments with potential customers. The learning acquired throughout this process should drive the entrepreneur to validate (or not) the original hypotheses. The simplicity of the lean startup approach, essentially based on experimentation, is quite appealing. This explains the enormous worldwide diffusion of this methodology, as well as the increasing attention that it has been receiving among practitioners and, quite recently, also among scholars (e.g., [11–13]). However, as the latest research has documented, the implementation of the lean startup approach is not without difficulties, which might possibly offset the advantages (e.g., [14–16]).

The aim of this chapter is to explain how the use of open-source hardware—which is essentially a set of open tools to support people, including potential entrepreneurs, who want to design their own hardware—might help reduce the costs usually associated with the lean startup experimentation approach. The chapter proceeds as follows. In the first section, we will describe the concept of lean entrepreneurship, pointing out not only its advantages and benefits but also its disadvantages and costs. In the second section, we will instead describe what open-source hardware is, focusing in particular on those characteristics that make open-source hardware applicable in a wide variety of settings by several different users—including potential entrepreneurs. Then, we will explain how the use of open-source hardware might actually help entrepreneurs to limit the possible costs usually associated with a lean startup approach, in order to maximize the net gains from experimentation.

2 The Rise of Lean Entrepreneurship

The concept of "lean" originates in manufacturing: more in detail, Womack et al. [17] have used the word "lean" to describe Toyota's production system, designed to address several key operational issues via the application of a few key principles, such as learning and continuous improvement of the supply chain. Ries [10] has tried to transplant the concept of "lean manufacturing"—and of the underlying key principles—into entrepreneurship, on the ground that "the theory that is the foundation of Toyota's success can be used to dramatically improve the speed at which startups find validated learning" (p. 188). Accordingly, the idea of Ries [10] is that, similarly to Toyota's manufacturing system, startup activity might take advantage of experimentation, learning, and continuous improvement.

In practice, the "lean startup method" (as depicted in Fig. 1) draws on the idea that startups should engage in a quasi-scientific experimentation for reducing the uncertainty about the viability of their business model. Entrepreneurs should

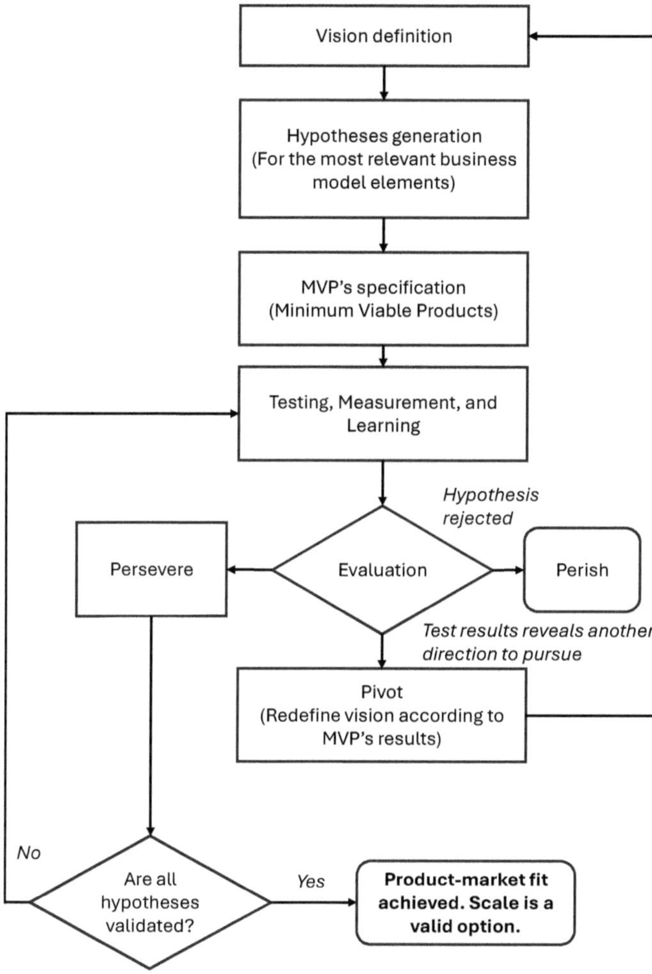

Fig. 1 The lean startup approach. Source: Eisenmann, Ries, and Dillard [18]

therefore generate explicit hypotheses about how they intend to create value with their products, and then should empirically corroborate such hypotheses. Such quasi-scientific experimentation passes through the creation of a "minimum viable product"—defined as an early version of the product, including all features needed to collect meaningful feedback about customers' actual preferences, and created at the minimum possible cost. The concept of minimum viable product is really at the core of the experimentation phase: it should be designed according to specific hypotheses about what customers like (or do not) and should be used to prove or disprove such hypotheses by collecting customers' feedback. In case potential customers appreciate the product, further iterations for refining the business idea are still possible and should be possibly conducted on a greater scale. In case, instead, the experiment does not support the initial expectations of the entrepreneurs—such that the business idea is not valuable as thought—a second key element of the startup methodology, or "pivoting," enters into action. In essence, pivoting represents a change in the trajectory of the development of the business idea, based on the knowledge obtained through the learning process. That change might involve one or more of the several dimensions of the business idea, such as, for instance, the targeted set of customers, the characteristics of the product itself, or how the value created by the product will be monetized.

Dropbox is usually described as the quintessential example of a successful business idea tested through a lean startup approach. Its founders, Drew Houston and Arash Ferdowsi, decided to test the quality of their idea by creating a 3-minute video of a product demo. Interestingly, this video was created even before having a working version of the product. In this sense, it is a compelling example of a minimum viable product: it was a representation of the Dropbox product and business idea, allowing for meaningful customer feedback, but at the minimum possible cost. More in detail, consistent with the lean startup methodology, this video tested the main hypothesis that customers were actually facing a sync problem when using the same file across different devices, which Dropbox could possibly solve. Potential users were not asked to use the actual product, but just to express their opinions on a video that represented the key aspects of the product. The feedback from potential customers allowed Dropbox to assess if there was demand for the product. Most importantly, this feedback provided a way for Dropbox to shape product development in line with consumer needs. Another example of a company that has successfully adopted a lean startup approach based on experimentation is provided by Murray and Tripsas [19]. They describe how Surface Logix, in commercializing soft lithography, was not sure whether to sell this technology as a product, as a service, or as a platform. Hence, the founders of Surface Logix embraced an experimentation approach by identifying the decision to be made (e.g., on which market should we focus), building explicit hypotheses (e.g., in which market our technology will be more profitable), and finally testing those hypotheses. This experimentation approach was quite crucial in shaping the direction and the ultimate success of the firm.

Based on the successful instances of companies that had adopted a lean startup methodology, as well as on the simplicity of its key principles, the lean startup approach nowadays constitutes a key theoretical foundation of many entrepreneurship courses and is ubiquitous across university-based entrepreneurship centers, incubators, and business schools [16]. Due to its success, the idea of lean startups has started attracting quite a lot of scholarly attention in order to better assess its effectiveness. For instance, Camusso et al. [13] show, in a field-experimental setting, that teaching entrepreneurs to think like scientists—by carefully framing their problems in terms of falsifiable hypotheses and crafting rigorous experiments for corroborating such hypothesis—lead them to avoid both false positives (or pursuing bad business ideas longer than they should) and false negatives (not pursuing business ideas that are instead promising). Similarly, [11] use a randomized controlled experiment to confront two types of approaches entrepreneurs can undertake when launching their own venture: a planning-based approach (based on adhering on a clear preexisting business idea) vs. an adaptive approach (based on keeping the flexibility to change the original ideas according to feedback). They document that the planning-based approach produces better outcomes in the short term; yet, in the long term, the adaptive approach performs better. Along similar lines, McDonald and Eisenhardt [12] in a qualitative study, done in the context of the social investing field, explain how successful new ventures essentially follow an experimentation logic, by testing their assumptions about value creation and keeping a substantial amount of organizational flexibility before committing to a certain business model.

However, previous literature, besides explaining the benefits possibly associated with the lean startup method, has also pointed out some costs related to the use of this approach. First of all, there is a cost inherent in creating a minimum viable product for experimentation. It is certainly true that, as noted by Kerr et al. [20], the cost of prototyping has declined dramatically in recent years, due to technological advancements. However, such costs have not decreased homogeneously across industries. The cost of prototyping drop has certainly been more salient in software and information technology businesses—where, indeed, as pointed out by Kerr at al. [20] firms that, ten years ago or so, would have costed 5 million dollars just to set up, nowadays can be created for less than 50 thousand dollars. This is due, for instance, to the diffusion of open-source software, which has lowered the costs associated with hiring programmers and/or writing software applications to be tested by potential users. However, in other industries where hardware is more relevant, the costs of prototyping are still high. Producing a minimum viable product, for testing, for instance, whether potential customers appreciate a new wearable device, might require significant investments for buying and assembling the necessary materials—and so it implies clearly higher costs compared to those of producing a new software application.

A second cost of experimentation is related to the difficulties of gathering information from customers. As pointed out by von Hippel [21], product developers—but also entrepreneurs who intend to launch a new product and collect feedback from customers—are confronted with the problem of collecting users' "sticky" information, or information that is costly to acquire, transfer, and use in a new location. The problem is that this information would be quite useful for developing a valuable product; yet, due to its stickiness, users are not able to transfer this information to product developers, and product developers are not able to gather it. Therefore, the traditional market research approach of gathering feedback for customers via simple surveys usually fails in understanding what customers really want, and a quite significant portion of the total variation in customer preferences might remain unaddressed [22]. The problem of not being able to capture exactly what users want might affect not only extant companies but also entrepreneurs when they present the minimum viable product in front of their potential customers and interact with them to gather feedback.

A third cost of experimentation is determined by the (low) quality of the feedback entrepreneurs can receive while interacting with customers. The lean startup approach points out the importance of getting customer feedback, grounded on the assumption that customers know what they might want. However, as argued by Felin et al. [16], this might not always be the case. Customers' assessments of a very innovative product might be negative, even if the product itself is valuable, just because customers are not used to similar products. Hence, engaging with customers might lead to the rejection of radical yet profitable ideas. What is worse, interacting with customers might even induce learning myopia and determine the inability to anticipate changes in customer preferences. Previous research has clearly shown that this type of learning myopia induced by customers is extremely dangerous for market leaders, and it might eventually lead to their disruption—as first documented by Christensen [23] in the context of the hard disk drive industry. However, there are valid reasons to argue that entrepreneurs might also be subject to the same learning myopia when presenting their minimum viable product to customers for receiving feedback. The result might be creating a product that, when it finally reaches the market, is already "old" and does not meet the customers' preferences.

A fourth cost of experimentation is associated with the lack of commitment and the ensuing disadvantages this lack might imply. Based on Ghemawat [24], commitment can be defined as an earlier choice that constrains later ones. So any choice, for representing a commitment, must be long-lasting and affect future decisions, in the sense that, once the choice is made, only a limited number of future courses of action are possible. This type of choice might bring relevant advantages, even in the context of a startup. For instance, as noted by Chavda [15], a startup committing to a certain business idea might not attract higher-quality collaborators or employees who prefer long-term contracts. Moreover, it might also direct the efforts of existing collaborators mainly to improve the final version of the idea, which is actually the most important one, rather than any intermediate—and so less important—version. However, the entire lean startup method is precisely based on the idea of avoiding any type of long-lasting early decisions. In doing so, the lean startup method, however, gives up any advantage that commitment might bring, in terms of finding valuable employees or collaborators and properly motivating them.

A final cost of experimentation is determined by possible misappropriation, occurring when entrepreneurs show their product to customers to gather feedback. The risk of misappropriation is a critical concern for firms in general (e.g., [25, 26]) and for early-stage ventures in particular (e.g., [27]). However, as explained and documented by Contigiani [14], in the context of launching a startup, this problem might be exacerbated by experimentation, as this essentially implies providing potential customers with information about a new idea—which obviously increases the risk of imitation. Whereas there are formal mechanisms of intellectual property rights for preventing imitation (such as, for instance, patents or copyrights), startups usually do not have the financial resources for leveraging them. At the same time, other informal mechanisms relying on the existence of complementary assets such as brand or reputation [28] are clearly not readily available for startups, as these assets usually take quite a lot of time to be developed.

After having explained the main challenges and costs associated with lean startup experimentations, in the next section, we will try to understand how and under what contingencies the use of specific technologies—"open-source hardware"—might contribute to enhancing the gains of experimentation while reducing its limitations.

However, before doing that, we will describe what open-source hardware is and how it works.

3 Open-Source Hardware and Its Use

The concept of "open-source" is thought to have its origins in the field of software. In particular, in 1984, Richard Stallman, who was at that time a researcher at MIT, created the Free Software Foundation (FSF) and launched the GNU project (GNU manifesto), which constitutes the roots of today's open-source movement. The idea behind the open-source movement is essentially that software, rather than being proprietary knowledge for commercial purposes, is knowledge to be freely distributed and developed by anyone interested in it. Accordingly, the terms of license of open-source software are designed such that software can be freely accessed, used, changed, and shared. Thanks to these terms of license, open-source software projects usually stem from a community of programmers, who, on a voluntary basis, develop, distribute, and support the software because they are themselves users of it. Linux is probably one of the most famous examples of open-source software. It was created in 1991 by Linus Torvalds, and it consists of a Unix-like operating system, for which the software code is free and available for anyone who wants to change and redistribute it. Linux has attracted a vast community of users all over the world, who, in a completely decentralized way, have contributed to the development of the Linux software. For this reason, Raymond [29] makes an interesting analogy between the Linux software development mode and a bazaar of different agendas and approaches. This bazaar governance mode ensured the success of Linux via user-driven design, rapid code improvement, and effective debugging.

The idea of openness characterizing the open-source software has quite recently been extended to hardware too, for defining open platforms that anyone can use and modify in order to design her or his own hardware. Arduino represents a clear instance of an open-source hardware platform (Fig. 2). It is a physical board that can be programmed to sense the external environment and interact with it by controlling physical actuators, such as lights and motors. The platform was originally developed for educational purposes at the Interaction Design Institute Ivrea in 2005 in order to engage students in artistic and design-oriented tasks. So, it was focused on simplicity, with the objective of involving a nontechnical audience. The team behind Arduino decided that the platform should be utilized via an open-source governance mode. Therefore, they protected Arduino under the Creative Commons license (e.g., [30]), establishing that anyone could produce copies, redesign it, or sell boards that copy the design, without the requirement of paying any kind of fee or royalty—the only requirement being that any derivative version must be used under the same license.

The success has been quite immediate: by 2006, Arduino had sold 5000 units, while the next year it sold 30 000 units and, until now, millions of copies have

Fig. 2 Picture of Arduino Uno board. Source: https://store.arduino.cc/arduino-uno-rev3

been sold worldwide. This vast diffusion of Arduino is also due, besides the relative simplicity of use, to the Creative Commons license. This has promoted the growth of a large community of users, which could share, reuse, and remix both codes and designs, in order to adapt the Arduino hardware for their goals. This community encompasses quite heterogeneous users, consistent with the original scope thought for the platform, which, as reported on the Arduino website, is "intended for artists, designers, hobbyists, and anyone interested in creating interactive objects or environments".[1] The wide reach of the Arduino community, in terms of users, is reflected in a broad range of applications of the hardware across different domains. Arduino has been in fact used by hobbyist inventors for producing several inventions, ranging from auto-lacing shoes, to a "thermometer scarf" with lights indicating the external temperature, to a garage door opener fingerprint scanner. However, companies have also been increasingly using Arduino: for instance, a firm called Botanicalls developed an Arduino-based device that can monitor house plants and send messages to apprise that they need to be watered.[2]

Following the example of Arduino, other open-hardware platforms have emerged. For instance, BITalino was created in 2013 with the goal of making the realization of devices able to capture and record bio-signals (such as electrocardiography, electromyography, electrodermal activity, and electroencephalography) available to everyone [31, 32]. BITalino is in fact a low-cost, open-source kit aimed at acquiring physiological signals in order to create wearables, self-monitoring apps, or biomedical devices. Notably, the devices realized by using BITalino have been shown to perform as well as conventional biomedical devices [31]. Moreover, the BITalino platform, besides being used in a wide range of biomedical applications, has also found successful applications in completely different fields such as, for instance, music composition (via a heart rate controller)[3] or drones controlled by gestures.[4]

[1] https://www.arduino.cc/en/Main/HomePage3

[2] Botanic-related projects integrating Arduino: https://www.botanicalls.com/kits/

[3] Music-related projects integrating BITalino: https://youtu.be/6PVBuPrA0pA and https://vimeo.com/136944173

[4] Drones-related projects integrating BITalino: https://youtu.be/GHrtN-rZHiE

The examples of Arduino or BITalino tell us about some quite important characteristics of open-source hardware. First of all, any open-source hardware, such as Arduino or BITalino, is essentially a toolkit, that is, an assembly of tools aimed at a particular purpose, which customers might readily use for designing and creating their own products [33]. These open-source toolkits are not only relatively easy to use for customers, but they are also quite cheap, because alternative versions of the same toolkits could be produced and sold by any company, relying on specifications that are completely public. For instance, a cheaper version of Arduino is Freeduino, designed from the same schematics as the original Arduino, but not branded by the Arduino company.

Second, open-source hardware platforms such as Arduino or BITalino are characterized by a high degree of generality. General technologies are technologies that can be applied to a wide variety of domains, with some—usually quite low—application costs [34]. There is quite a lot of variation among technologies in the degree of generality they exhibit. Some technologies can only be used in quite specific settings. Others are instead so flexible that they can be defined as "general purpose technologies," due to their capacity to perform a generic function that is central to the functioning of an extremely large number of products and production systems [35]. To be sure, only a few technologies can reach a pervasiveness and wide applicability comparable to that of the steam engine or electricity. Yet, several technologies, even without being general purpose, are still general enough to be applied to a broad set of domains, which is often beneficial for both customers and suppliers, in the former case, due to the adaptability of a general technology to a large variety of needs [34] and, in the latter case, due the operational and strategic flexibility a general technology could provide (e.g., [36–38]). In this respect, any open-source hardware is a platform, which, by design, is applicable, with some modifications, across a wide range of different domains. So, open-source hardware can certainly be defined as a general technology.

A final important characteristic of open-source hardware is the existence of a community of users and developers around it. Open-source hardware platforms such as Arduino or BITalino have tended to grow and develop thanks to innovation communities whose members not only share instructions and codes but also provide feedback for new products and applications. As Von Hippel [33] puts it, innovation communities are organizational forms where members cooperate in the development, testing, and diffusion of user-initiated innovations, sharing knowledge, information, and occasionally resources [39]. These communities are mainly composed of users that derive direct benefits from an artifact, because they enjoy creating or using it [40]—even if sometimes innovation communities might encompass not only users but also manufacturers. In addition, participation in innovation communities is usually voluntary, which implies no remuneration from the community. Voluntary participation is also explained by the fact that innovation communities are not just functional but are also social communities, in that members experience a sense of belonging and social identity—which might work as a substitute for monetary rewards.

Interestingly, the characteristics of open-source hardware we have underlined—being an (affordable) toolkit, being a general technology, and relying on a large community of users—are strictly intertwined and reinforce each other. On one side, any open-source hardware platform is designed as a low-cost and general toolkit precisely in order to attract a large variety of users with different needs, who might become members of the innovation community flourishing around the platform. On the other side, the existence of a large innovation community of users with diverse needs and skills enhances the process of discovering new applications in different fields, which in turn strengthens the generality of the platform.

4 How Open-Source Hardware Might Reduce the Costs of Lean Entrepreneurship

After having clarified the characteristics of open hardware platforms, we are able to explain how the use of these platforms in the context of a lean startup approach might actually reduce the costs usually associated to experimentation (Fig. 3)—as summarized in Table 1—and possibly enhance the strengths of a lean startup approach.

The first lean startup experimentation cost we have considered is the cost of creating a minimum viable product for testing a novel product. In this respect, as made evident by the example of Arduino and its broad applications, open-source hardware might offer an affordable alternative for designing new devices, thanks to its generality, which allows for applications in a wide range of domains, and the existence of a supportive innovation community, which reduces the costs of

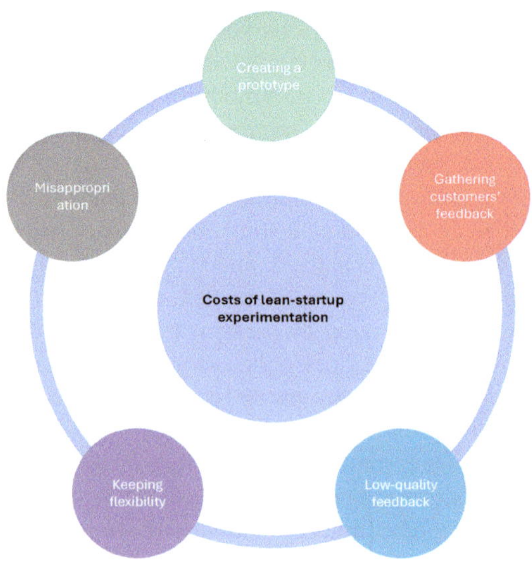

Fig. 3 The costs of lean startup experimentation

Table 1 Costs of lean startup experimentation and solutions provided by open-source hardware

Cost	Explanation	Solution
Cost of creating a prototype	In the case of creating a hardware minimum viable product, the costs of buying and assembling materials together might be substantial	Open-source hardware offers a cheap alternative for designing new devices in a wide range of domains, due to its generality
Cost of gathering customers' feedback	Some customers' information is sticky: users are not able to transfer it to entrepreneurs, and entrepreneurs are not able to gather it. So the final version of the product might not correspond to what customers really want	Open-source hardware is a general toolkit that customers can use to embed their feedback directly into the product, possibly enhancing the likelihood of pivoting
Cost of low-quality feedback	Engaging with customers might induce learning myopia, that is, the inability to anticipate changes in customer preferences	The community developing open-source hardware is populated by "lead users," who usually anticipate future preferences of the market
Cost of keeping flexibility	The lean startup method, with its focus on avoiding any type of long-lasting early decisions, gives up any advantage commitment might bring, in terms of finding valuable employees/collaborators and properly motivating them	The community developing open-source hardware is composed by members who are intrinsically motivated to support the creation of new products based on that hardware
Cost of misappropriation	Lean startup experimentation implies giving the market access to information about an idea, which obviously increases the risk of imitation	The possibility of reaching an already existing wide community of potential users might help entrepreneurs to establish their brand and reputation in a quite short period of time

designing a prototype. Interestingly, affordable and general open-source hardware not only reduce the cost of prototyping but also promotes the likelihood of pivoting. Destroying an affordable device is economically and psychologically easier than doing the same with a costly device, especially when its components can be readily reassembled to create a completely new device. Hence, open-source hardware platforms, being both affordable and general, encourage entrepreneurs to change—even radically—their product idea when customers do not seem to like it.

A second lean startup experimentation cost derives from gathering feedback from users. As we argued, due to information stickiness, it might be hard or very costly for entrepreneurs to understand customers' needs and preferences. However, toolkits such as the Arduino might eliminate the costly exchange of need-related information between users and entrepreneurs during the process of testing a new product. Especially when open-source hardware is user-friendly, final customers might directly embed their feedback and preferences into the product they are testing—

rather than communicating them to entrepreneurs [41]. Furthermore, thanks to the generality of open-source hardware platforms, the feedback incorporated by customers into the prototype, rather than being just incremental, might determine a pivoting: the product might develop into a completely different direction than what was originally thought, for instance, leading to a shift in the targeted market.

A third cost derives from the challenge of collecting high-quality feedback and the losses implied by low-quality feedback—in terms of the inability to meet future customers' preferences. Interacting with innovation communities developing open-source hardware might alleviate this cost, since these communities are largely populated by lead users—or users who first face needs that will become salient for the rest of the customers only months or even years later [42]. Lead users are likely to join innovation communities developing open-source hardware, precisely because the current market offer does not satisfy their needs, and so they are pushed to find a solution to these needs elsewhere. Potential entrepreneurs using open-source hardware for constructing prototypes might therefore get in contact with lead users and—thanks to the feedback coming from these users—be able to anticipate customers' future preferences.

A fourth cost associated with the lean startup experimentation method is associated with a lack of commitment. Until entrepreneurs do not commit to a certain business idea, they might find it difficult to find valid collaborators for developing their product, because normally, these would prefer long-term contracts. They might also face difficulties in providing the right motivations to employees, who might put too much effort into the intermediate versions of the product, at the detriment of the effort they will put into the final version. However, the communities developing open-source hardware are composed of members who are intrinsically motivated to support the creation of new products based on that hardware. Their members usually provide strong support to any new projects on a completely voluntary basis, without the need for any particular contractual devices for raising their incentives.

The final cost that entrepreneurs adopting a lean startup experimentation approach have to consider is related to the risk of misappropriation. Gathering feedback from customers about a new business idea naturally enhances the likelihood of that idea being imitated. As we argued, startup companies usually do not have the resources necessary for using formal intellectual property rights. Furthermore, they also lack those complementary assets, such as brand and reputation, which would make their business less imitable. However, using open-source hardware opens the possibility of reaching an already existing wide community of potential users. An entrepreneur capable of leveraging the social ties linking innovation community members to each other, might—for instance, through online word of mouth—accelerate the development of a brand and reputation among future customers, so protecting the business idea.

5 Summary

The lean startup approach certainly has relevant merits, which explain its enormous diffusion. It is a relatively simple methodology, easy to teach, and readily applicable by entrepreneurs for reducing the uncertainty normally characterizing entrepreneurial activity, especially at the earlier stages. Yet this approach is also fraught with problems and risks. In this chapter, we have shown how the use of open-source hardware might represent a viable solution for effectively addressing some of the challenges usually associated with the implementation of a lean startup approach. Indeed, some firms have already been using open-source hardware for experimenting with their new products. One example is represented by De La Rosa Research, a Colombian biomedical company, which wanted to create a highly customizable system for bio-signal data acquisition, able to operate in extremely difficult environmental conditions, such as underwater, in jungles, or even in deserts. In order to reduce development costs and accelerate production, the company used BITalino: the result of this effort has been Daqphibian,[5] a device that can be used for research in a variety of different fields (e.g., military research, aquatic research, or human-computer interface research). A further example is Diabeto, an Indian startup producing a device that plugs into a glucometer and transfers the information to a smartphone app. The first prototype of the device was originally built and designed based on an Arduino board. As Shreekant Pawar, the founder of Diabeto, says: "open-source hardware lets you cut down on your prototyping and design time significantly and the vast community support helps you at every stage".[6] Drawing on these examples, we hope this chapter might stimulate other potential entrepreneurs to consider the possibility of using open-source hardware platforms when launching their new ventures.

References

1. R. Cantillon, Essai sur la nature du commerce en général, edited and with an English translation by Henry Higgs, 1755. Brewer Edition, Essay on the Nature of Commerce in General (2001)
2. J.-B. Say, *Traité d'économie Politique, Ou Simple Exposition de La Manière Dont Se Forment, Se Distribuent et Se Consomment Les Richesses. Tome 3/; Cinquième Édition, Augmentée d'un Volume et à Laquelle Se Trouvent Joints Un Épitome Des Principes Fondamentaux de l'économie Politique et Un Index Raisonné Des Matières ; Par Jean-Baptiste Say,...* 1826
3. F. Knight, *Risk, Uncertainty, and Profit* (Hart, Schaffner & Marx; Houghton Mifflin, Boston, 1921)
4. A. Marshall, *Principles of Economics*, 8th edn. (Macmillan, London 1930).
5. P. Drucker, Entrepreneurship in business enterprise. J Bus Policy **1**, 3–12 (1970)

[5] http://www.delarosaresearch.com/indexdaq.php

[6] https://www.gadgetsnow.com/more-gadgets/Indian-creates-Google-Glass-clone-for-Rs-4500/articleshow/40486647.cms

6. F. Scherer, D. Harhoff, Technology policy for a world of skew-distributed outcomes. Res Policy **29**, 559–566 (2000)
7. S. Shane, Why encouraging more people to become entrepreneurs is bad public policy. Small Bus. Econ. **33**, 141–149 (2009)
8. A. Dixit, R. Pindyck, *Investment Under Uncertainty* (Princeton University Press, Princeton, 1994)
9. A. Contigiani, D. Levinthal, Situating the construct of lean start-up: adjacent conversations and possible future directions. Ind. Corp. Change **28**, 551–564 (2019)
10. E. Ries, *The Lean Startup: How Today's Entrepreneurs Use Continuous Innovation to Create Radically Successful Businesses* (Crown Business, New York, 2011)
11. C. Eesley, L. Wu, For startups, adaptability and mentor network diversity can be pivotal: evidence from a randomized experiment on a MOOC platform. MIS Q. **44**, 661–697 (2020)
12. R. McDonald, K. Eisenhardt, Parallel play: startups, nascent markets, and effective business-model design. Admin. Sci. Q. **65**(2), 483–523 (2020)
13. A. Camuffo, A. Cordova, A. Gambardella, C. Spina, A scientific approach to entrepreneurial decision making: evidence from a randomized control trial. Manag. Sci. **66**, 564–586 (2020)
14. A. Contigiani, Experimentation and appropriability in early-stage ventures: evidence from the us software industry. SSRN Electron. J. **44**(9), 2128–2174 (2023)
15. A. Chavda, The downside of experimentation: Evidence from television shows (2018)
16. T. Felin, A. Gambardella, S. Stern, T. Zenger, Lean startup and the business model: experimentation revisited. Long Range Plan. **53**(4), 101889 (2019)
17. J. Womack, D. Jones, D. Roos, *The Machine That Changed the World* (Rawson Associates Scribner Simon & Schuster, New York, 1990)
18. T. Eisenmann, R.E, S. Dillard, Hypothesis-driven entrepreneurship: the lean startup, 2011. Harvard Business School Entrepreneurial Management Case No. 812-095
19. F. Murray, M. Tripsas, The exploratory processes of entrepreneurial firms: the role of purposeful experimentation, in *Business Strategy over the Industry Lifecycle*, vol. 21 (Emerald Group Publishing Limited, pp. 45–75, 2004)
20. W. Kerr, R. Nanda, M. Rhodes-Kropf, Entrepreneurship as experimentation. J. Econ. Perspect. **28**, 25–48 (2014)
21. E. Hippel Sticky information and the locus of problem solving: implications for innovation. Manag. Sci. **40**, 429–439 (1994)
22. E. Hippel, R. Katz, Shifting innovation to users via toolkits. Manag. Sci. **48**, 821–833 (2002)
23. C. Christensen, *The Innovator's Dilemma: When New Technologies Cause Great Firms to Fail* (Harvard Business School Press, Boston, 1997)
24. P. Ghemawat, Market incumbency and technological inertia. Mark Sci. **10**, 161–171 (1991)
25. R. Conti, Do non-competition agreements lead firms to pursue risky R&D projects? Strateg. Manag. J. **35**, 1230–1248 (2014)
26. F. Castellaneta, R. Conti, A. Kacperczyk, Money secrets: how does trade secret legal protection affect firm market value? Evidence from the uniform trade secret act. Strateg. Manag. J. **38**, 834–853 (2017)
27. F. Castellaneta, R. Conti, F. Veloso, C. Kemeny, The effect of trade secret legal protection on venture capital investments: evidence from the inevitable disclosure doctrine. J. Bus. Ventur. **31**, 524–541 (2016)
28. D. Teece, Profiting from technological innovation: implications for integration, collaboration, licensing and public policy. Res. Policy **15**, 285–305 (1986)
29. E. Raymond, *The Cathedral and the Bazaar: Musings on Linux and Open Source by an Accidental Revolutionary* (O'Reilly, Cambridge, 1999)
30. Y. Lin, T. Ko, T. Chuang, K. Lin, Open source licenses and the creative commons framework: license selection and comparison. J. Inform. Sci. Eng. **22**(1), 1–17 (2006)
31. D. Batista, H. Silva, A. Fred, C. Moreira, M. Reis, F.H.A, Benchmarking of the bitalino biomedical toolkit against an established gold standard. Healthcare Technol. Lett. **6**(2), 32–36 (2019)

32. H. Silva, F.A, R. Martins, Biosignals for everyone. IEEE Pervasive Comput. **13**(4), 64–71 (2014)
33. H.E, Democratizing innovation: The evolving phenomenon of user innovation. J. fur Betriebswirtschaft **55**, 63–78 (2005)
34. T. Bresnahan, A. Gambardella, The division of inventive labor and the extent of the market, in *General Purpose Technologies and Economic Growth*, ed. by E. Helpman (The MIT Press, Cambridge, 1998), pp. 253–281
35. N. Rosenberg, *Inside the Black Box: Technology and Economics* (Cambridge University Press, Cambridge and New York, 1982)
36. A. Gambardella, A. McGahan, Business-model innovation: general purpose technologies and their implications for industry structure. Long Range Plan. **43**, 262–271 (2010)
37. R. Conti, A. Gambardella, E. Novelli, Specializing in general purpose technologies as a firm long-term strategy. Ind. Corp. Change **28**, 351–364 (2019)
38. R. Conti, A. Gambardella, E. Novelli, Specializing in generality: firm strategies when intermediate markets work. Organ. Sci. **30**, 126–150 (2019)
39. N. Franke, S. Shah, How community matters for user innovation: an exploration of assistance and sharing among end-users. Res. Policy **32**, 157–178 (2003)
40. H.E, *The Sources of Innovation* (Oxford University Press, New York, 1988)
41. N. Franke, F. Piller, Value creation by toolkits for user innovation and design: the case of the watch market. J. Prod. Innov. Manag. **21**, 401–415 (2004)
42. E. Hippel, Lead users: a source of novel product concepts. Manag. Sci. **32**, 791–805 (1986)

Technology Transfer: From Research to Industrialization

Aline Santos Silva, Ana Sofia Cacais do Carmo, and Hugo Plácido da Silva

1 The Four Phases of the Industrialization Process

Research is a continuous process of putting ideas to the test, and the transfer of technology (from research to industrialization) should be an organic and necessary process. However, regardless of both technological and research entities and industry having the same objectives, their communication approaches are often different, which creates a separation between them. The key is understanding how to be the bridge between the research language and the industry so that the communication is fluid; if we understand each step in the industrialization process, the communication will be easier.

The industrialization process can be divided into four main phases, which are described in the following sections.

Phase 0: Project Review
This is the first step of interaction between industry and research; it is known as Phase 0 because it is the pre-development phase. This is when it is pre-established and questioned whether there is a market need and if the project really is valid and promising enough to go to market. At this point, it is necessary to think about what

A. S. Silva (✉)
INESC TEC - Institute for Systems and Computer Engineering, Technology and Science, Porto, Portugal

IT - Instituto de Telecomunicações, Lisbon, Portugal
e-mail: aline.s.silva@inesctec.pt

A. S. C. do Carmo · H. P. da Silva
Instituto Superior Técnico, Universidade de Lisboa, Lisbon, Portugal

IT - Instituto de Telecomunicações, Lisbon, Portugal

H. P. da Silva et al. (eds.), *Open Source Biomedical Engineering*,
https://doi.org/10.1007/978-3-032-03655-1_9

makes a product or service unique, what makes it better than the competition. To do this, one should try to have the answers to the following questions:

- What is my product or service, and what does it do?
- How does my product or service solve a certain problem/market need?
- How is it different from other products and/or services?
- Who will buy it? Who will be my clients?
- Why will they buy it?
- How will the product or service be promoted and sold/offered?
- What kind of technologies would be necessary to develop my product/service?

Once these questions are answered, a more descriptive process can be started about the product, including a state-of-the-art review, intended use, system requirements, and Proof of Concept (PoC) reports. During the process of creating and testing the PoC, everything must be documented and described, both the positive and negative results, in order to understand this new technology and the parameters that restrict it. This phase is what will define whether the technology will be industrialized or not.

Phase 1: Development and Pre-Project

In this phase, the product itself is developed, for which the development plan for the completion of each task will need to be established. For medical devices, this plan will include a field for each task, along with the corresponding deadline, resources (human, infrastructure, and monetary), person responsible, and an additional field for observations. This planning must be followed up by the project manager whenever there are changes in dates that affect the project's deadline. Moreover, it is at this stage that three crucial tasks should be addressed: collection of product requirements, risk analysis, and competition analysis (Fig. 1 illustrates the main aspects that should be addressed in Phase 1 of certification).

After the planning is done, the creation of prototypes for the PoC begins. The PoC is nothing more than the first physical prototype of the product, which makes it possible to visualize the advantages and disadvantages of this first version of the product.

With a clearer perception of the limitations of the product, an explicit set of requirements, and a better understanding of the system after the PoC, the next step is to elaborate the Focus Matrix, the Failure Mode and Effects Analysis (FMEA) and the Risk Analysis Matrix (RAM) [2]. The former Risk Report (RR) is the description of the whole risk management process; it is based on the Focus Matrix, the FMEA and the RAM, where it details general risks (Focus Matrix), risks to the functionality of the device (FMEA), and the estimation of the degree of occurrence of the risk (RAM) [2, 3]. On the other hand, the FMEA identifies and assesses failure modes and their effects on the functionality of the device, its usability, software, or even manufacturing. Contrary to the FMEA, the RAM examines the possibility and probability of a hazardous situation arising, which may lead to harm to the patient, user, or the environment. However, keep in mind that none of these three analyses live in isolation, as they are crucial complements to each

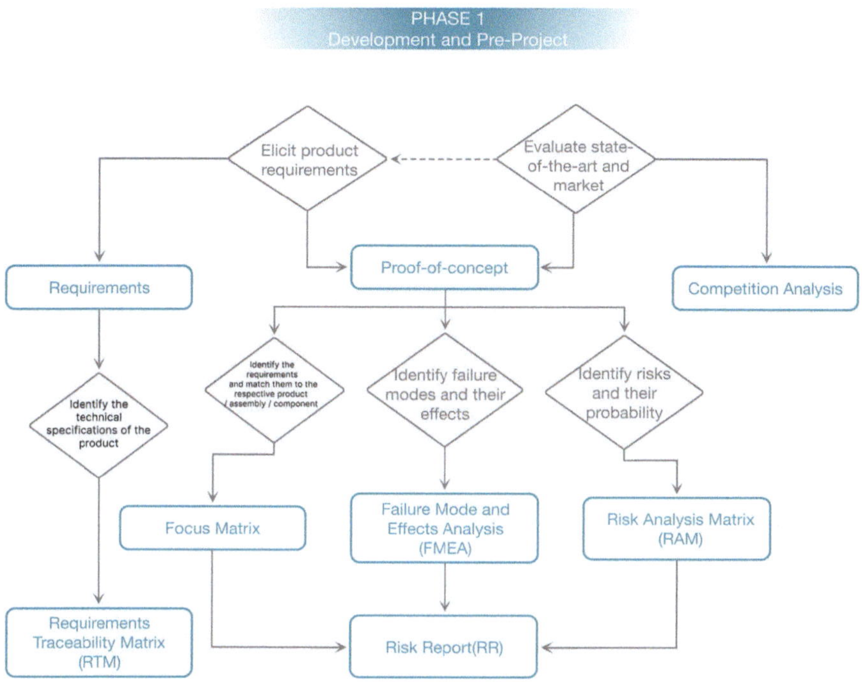

Fig. 1 Review of the processes of the development and pre-project phase [1]

other (for example, a failure in device functionality may lead to an actual hazardous situation for the user). Finally, these three documents will be used to write the RR, which entails a comprehensive description of the processes that justify and indicate suggestions for each possible risk influencing the device's operation and user safety. It comprises one of the three most important documents of a product certification (along with the Technical Documentation and Clinical Evaluation Report (CER)) [4]. Figure 2 illustrates the main points that should be addressed in the RR according to Regulation (EU) 2017/745, ISO 24971, ISO 13485, and ISO 14971 [4, 5].

In addition to the RR, this phase of the project will also start the preparation of the Requirements Traceability Matrix (RTM), which is a matrix that makes explicit the direct relationship of requirements to each other or to the other components of the project. In order to facilitate the process of prototype approval and improvement for the PoC.

It is also at this stage that we start building the Technical Documentation for medical device certification, which is described in Sect. 2.

Phase 2: Preparation for Industrialization

Preparation for industrialization begins with a review of the final product by the project team. In this phase, all members of the team are responsible for determining in consensus whether the project will proceed or not. If the project is to proceed, it

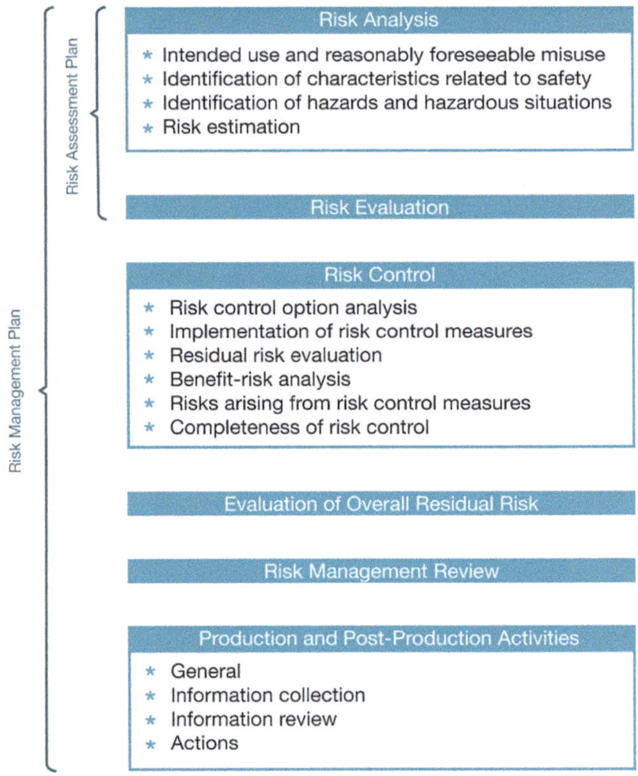

Fig. 2 A schematic representation of the risk management process [5]

is in this phase that the documentation of the PoC tests must be finalized and the final product design defined, or its realization justified in Phase 3.

In addition to the approval of the manufacturing and assembly process, the quality control process must be approved by comparison with the performance of reference devices validated by medical device literature or certification. Figure 3 shows the flowchart of the process for device approval.

At this stage, one must contact a Food and Drug Administration (FDA)—accredited issuing agency to assign a Unique Device Identifier (UDI) to the product, which is a unique numeric or alphanumeric code related to a medical device. Moreover, through the European Commission,[1] the European Medical Device Nomenclature (EMDN) of the device should be determined. And finally, Technical Documentation (Sect. 2) must be submitted to the body that will certify the medical product, for example, to a European Database for Medical Devices (EUDAMED) [6, 7].

[1] https://webgate.ec.europa.eu/dyna2/emdn/Z1206

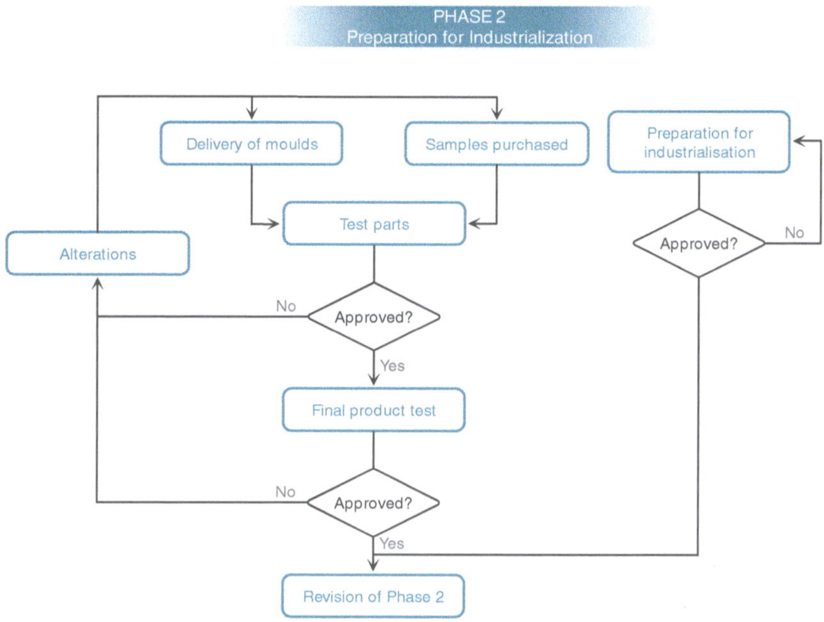

Fig. 3 Review of the processes of the preparation phase for industrialization (based on [1])

Phase 3: Pre-Series and Project Validation

Each company has a date for Final Delivery, as defined during planning. This delivery is nothing more than the date that the product is legally ready to go to market as a medical device. Many companies consider the Final Delivery to be the closure. But this stage has many more elements, which are relevant not only to the project itself, but also to the organization's next decision-making. At the end, it is recommended to do an overall review of the project to ensure that everything that was planned was done and that no elements were lost in the middle of the process. In addition, it is of added value to analyze the successes and mistakes, as lessons can be learned and processes improved, allowing more assertive executions in the future.

At this stage, the certification process should have been completed with the recognition of the device as a medical device by the certification bodies. After product validation, a set number of devices must be produced and validated internally to verify compliance with the requirements and validate the functioning of the product. This analysis should include, where appropriate, planning for post-marketing or post-implementation outcome evaluation arrangements, nonconformities, complaints, and stakeholder perception and impact assessments. Figure 4 shows the flowchart for the processes of the pre-series phase and project validation.

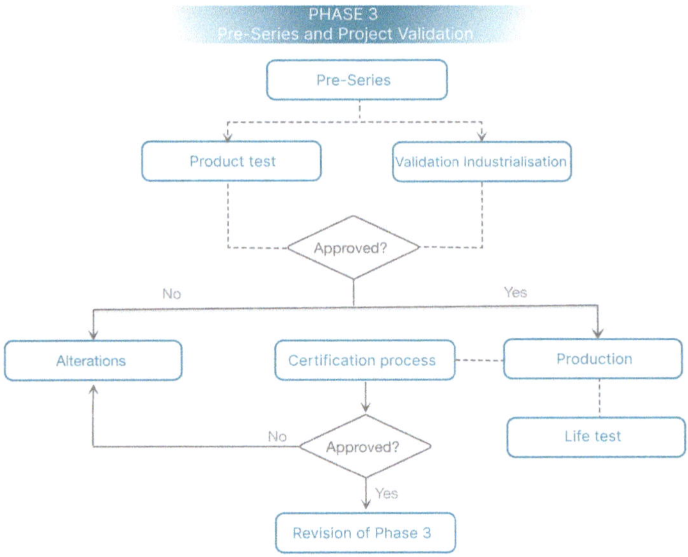

Fig. 4 Review of the processes of the pre-series phase and project validation [1]

2 Technical Documentation for Medical Device Certification

The global medical device market is governed by a wide range of national and international standards and regulations and medical device certification standards. These regulatory requirements are complex and vary between regions, which can make it difficult to obtain medical approval for the products in the target market. Technical Documentation provides information about the design, manufacture, and operation of a product and must contain all the information necessary to show that the product meets the applicable requirements [8]. To place a product on the market, the manufacturer must follow certain rules, namely:

- Prepare the Technical Documentation before placing the product on the market.
- Ensure that the Technical Documentation is available for consultation by the market surveillance authorities (if they ask to see it) as soon as the product is placed on the market.
- Keep the Technical Documentation for 10 years from the date the product was placed on the market (unless explicitly stated otherwise).

The Technical Documentation consists of 4 main sections: Introduction; Administrative Information; Technical Information; and Annex List.

Section 1: Introduction
This section is defined by the product scope, i.e., the reason for making the medical device, as well as the type of device, the manufacturer, and the regulations that

will cover that device, which, for regular medical devices, is the Regulation (EU) 2017/745 [9].

Section 2: Administrative Information
Manufacturer information and subcontractors (which are all external companies that have a contract and supply a specific part to the manufacturer) should be described in this section. The following shall be indicated: the legislation applied to the device; and the medical device identification codes. These identification codes consist of 4 types—the commercial reference of the device, the EMDN, and the UDI—and in addition to these, each commercially distributed device will still have an additional code that is automatically generated for the identification of a single device (product ID). This section must also include the class of the medical device, along with the rationale for the chosen class, indicating the rules that are associated according to Regulation (EU) 2017/745—Annex VIII [9].

Section 3: Technical Information
This section is subdivided into several topics that are further described in Sect. 3 of this chapter. Each topic addresses different aspects that constitute the entire industrialization process, from its motivation to the lifetime of the device, which are based on the guidelines established by Regulation (EU) 2017/745—Annex II [9]. This information should be described succinctly and directly so that the content is easy to understand.

Section 4: Annex List
In order to prevent the Technical Documentation from becoming too large, it is recommended to include some of its topics as separate documents in the annexes. This should be applied to extensive topics, such as the RR and CER. Another document that should be included as annex is the one that addresses aspects of the research and comparison of competing technologies to the developed device.

In the certification process, the Technical Documentation must indicate all the annexes that are required for each section. So that the Technical Documentation is the main document where all the documents that will be used for the certification process are identified.

3 A Deep Dive into the Technical Information

This section describes all the information that should be covered in the Technical Information section.

3.1 Medical Device Description

The goal of this section is to describe the device, from its conception to its functionality. Ideally, the segmented product design should be illustrated with 3D drawings, indicating each part with its respective quantity, reference code, and functionality description. It is also recommended to present the dimensions and final aesthetics of the product. If the device has additional platforms beyond the physical product, such as apps and websites, these should also be briefly described. This description should provide a comprehensive summary of the device and its composing parts.

3.2 Intended Use

This contains the manufacturers' main objectives related to the purpose of the product, process, or service. This will include a description of the target population, the intended user, the environment of use, contraindications, warnings, and product claims to ensure that the purpose of the device is achieved.

3.3 State of the Art

In order to better understand the need for the creation of the product, the state of the art tells the story and the reason that led to the development of the final product. It is in this space that a large part of the questions proposed in section "Phase 0: Project Review" will be answered. One should detail what caused this need and the technologies that are found today to satisfy it. In short, the state-of-the-art refers to the highest level of overall development of a device, technique, or scientific field reached at a given time. However, in some contexts, it can also refer to a level of development reached at a particular time as a result of the common methodologies employed at the time.

3.4 Sales, Complaints, and Post-Market Surveillance

Post-market Surveillance (PMS) refers to the requirement that manufacturers monitor after the medical devices are approved for sale and are used in the marketplace. PMS is a regulatory requirement in significant markets, including the European Union (EU) and the United States (USA). Traditionally, PMS relies on reactive data collection [10]. Manufacturers collect and report adverse events from a device post-marketing. These adverse events are either reported directly to the manufacturer or may be reported to regulatory agencies. Post-marketing monitoring is critical to the improvement and evolution of the innovation process. Therefore, it will be necessary to evaluate annually, or if new information is identified from post-market monitoring that has the potential to change the current evaluation. In addition, an evaluation of the project results should be performed biannually for 3 years after project validation. For innovative devices, it should be indicated that since it is a new device that has not yet been commercialized, there are no records of sales, complaints, or PMS activities [11]. Collecting quality, performance, and safety data over the lifetime of a device helps manufacturers create a complete risk/benefit profile for their device and course correct quickly if problems arise.

3.5 EU Declaration

It is a mandatory document that the manufacturer must sign to declare that the products conform to EU requirements. By signing, the manufacturer assumes full responsibility for the conformity of the product to applicable EU legislation. This document must contain information on the manufacturer, the product, and its traceability forms, such as the declaration of responsibility and the relevant legislation with which the product complies [12].

3.6 Technical Standards and Common Specifications

This part describes all standards and regulations needed during the entire device development process. These are standards or requirements set for a repeatable technical task, thus allowing them to be performed according to a standard protocol [12].

3.7 General Safety and Performance Requirements

General Safety and Performance Requirements (GSPR) are the requirements that devices have to meet, according to their scope and intended use. The performance of the devices must prove how well they can perform with minimal clinical risk to the health of the patient or user. Device performance must not be adversely affected by transportation and storage. Generally, an appendix is created, which describes the GSPR according to Annex I—Regulation (EU) 2017/745 [9]. In addition, it should indicate the standards that were followed for the realization of each requirement [13].

3.8 Manufacturing Process and Subcontractors

In this section, the manufacturing process should be described step by step, along with its subcontractors and the main tests on the assembly line. In addition, it is necessary to have subsections that have the description and control of the raw materials, traceability procedures, and production control.

3.9 Information for Users

The product ID provides a unique product identification with a serial number. The label is printed with the model and product ID and a separate QR code label; this is attached to the device, thus allowing faster system connection for initial setup. The type of certification it has and the manufacturer's information are also provided. In addition, there is the user manual, which is one of the most important documents to develop. It should be easy to understand and detailed with all the important aspects that the user should know about the product, from how to install it, requirements for a good operation, how to use it, possible errors and warnings, along with how to solve them, recommendations, and suggestions. The manual should explain each step in detail with great care. If assembly of the product is considered complex, it may be advisable to include pictures and detailed instructions in the manual.

3.10 Design, Validation, and Verification

Any aspect that adds to the device and makes it unique should be described in this part, including any type of external accessory, both physical and digital, such as a platform for user interaction. The specific validations that justify the design and performance of the product should be addressed. In general, Design Verification is the confirmation by objective evidence that the design result matches the design input. Design validation means establishing by objective evidence that the specifications (detailed requirements) are in accordance with user needs and intended use(s).

3.11 Risk Management

Risk management is the process of identifying, evaluating, and controlling the possible risks one may have with the device. These risks stem from a variety of sources, including financial uncertainties, legal liabilities, technology issues, strategic management errors, accidents, and natural disasters [3].

Successful risk management helps the manufacturer consider the full range of risks faced. Risk management also examines the relationship between the risks and the cascading impact they can have on product performance. For medical devices, there are a number of regulations that control the risks that the device may have, so this document should be described based on them:

- ISO 14971 [14]: Medical devices—application of risk management to medical devices
- ISO 13485 [15]: Medical devices—quality management systems, requirements for regulatory purposes
- ISO 24971 [16]: Medical devices—guidance on the application of ISO 14971
- Regulation (EU) 2017/745 [9]: Regulation (EU) 2017/745 of the European Parliament and of the Council on medical devices

Besides detailing all possible risks, the answers to some questions should also be provided, such as:

- What is the type of benefit the medical device offers to patients?
- What is the magnitude of the benefit?
- What is the probability of the benefit?
- What is the duration of the benefit?
- Are there any benefits for the healthcare professionals or caregivers?

3.12 Clinical Evaluation

As the medical devices being evaluated are usually produced using conventional processes and using raw materials with established use in the market, the functional and safety performance under normal conditions of use is already well established [17].

Meanwhile, the clinical evaluation is performed on the basis of clinical performance in accordance with Art. 61 of Regulation (EU) 2017/745. To this end, a CER must be developed, encompassing all clinical data regarding the device, whether derived from clinical trials or literature research. Only for some devices are clinical trials required (and mandatory), and if they are conducted, this information must be included in the CER. For those products for which clinical trials are not mandatory, the CER can be obtained by describing similar products.

As a general rule, class I devices are exempt from clinical testing. However, whether a device needs to undergo clinical trials will depend on its class, whether it is substantially similar to or based on an already approved device, and/or whether or not significant clinical data to support the new device has already been published in medical literature. In the case of a device that is new to the market and does not have significant medical literature, it is suggested that clinical trials be done first for this report.

It is worth noting that clinical trials must be authorized by an ethics committee and by a regulatory body such as Infarmed.

3.13 Clinical Performance and Safety Summary

The Summary of Safety and Clinical Performance (SSCP) is intended to provide public access to an up-to-date summary of clinical data and other information about the safety and clinical performance of the device. The SSCP will be an important source of information for intended users—both healthcare professionals and, if relevant, for patients. It is one of several means intended to fulfill the objectives of the Medical Devices Regulation (MDR) to increase transparency and provide appropriate access to information. SSCP differs from clinical evaluation in that it is the clinical investigation of the product after it has been placed on the market.

3.14 Post-Market Clinical Follow-Up

Market surveillance activities are intended to collect post-marketing device data that can be used as inputs for risk management, clinical evaluation, or product design and development. These activities are described in the PMS plan, and the results are compiled, analyzed, and evaluated in the periodic safety update report.

3.15 Biological Safety

Biological safety refers to the use of specific practices, safety equipment, and specially designed buildings to ensure that workers, the community, and the environment are protected from accidental exposure or unintentional release of infectious agents, toxins, and other biological hazards. Biological safety assessments should be in accordance with ISO 10993-1 [18]. The biological evaluation of any medical material or device intended for use in humans should form part of a structured biological assessment plan within a risk management process in accordance with ISO 14971, Annex I [14]. This risk management process involves identifying the biological risks and determining their acceptability. The biological assessment must be planned, performed, and documented by knowledgeable and experienced professionals.

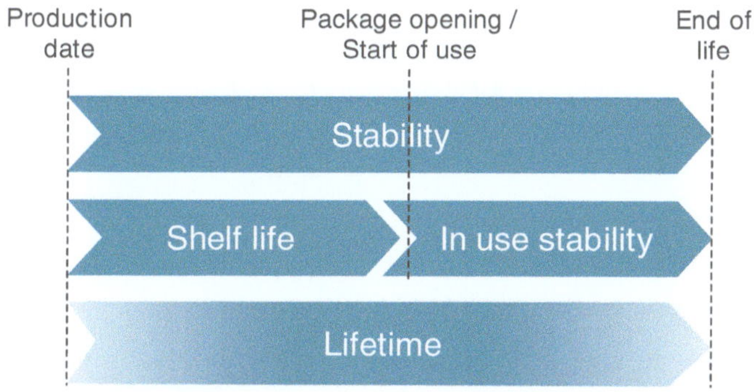

Fig. 5 Stages of a device's useful life

3.16 Packaging

It should include a description of the various levels of packaging, as well as tests to determine that the device is packaged with the greatest possible protection so that it is not damaged during transport to its final destination.

3.17 Expiration and Stability Tests

This is one of the last tests to be done after the manufacturing and development process is complete. In these tests, the following aspects must be checked:

- **Stability**: Property of the device indicating that the device remains safe and functions as intended throughout its life cycle
- **Shelf life**: Recommendation of time that products can be stored, during which the defined quality remains acceptable under the expected (or specified) conditions of distribution, storage, and exposure
- **In use stability**: Time between putting into service/opening the package and the end of life of the device

The determination of these aspects is extremely important as it is directly related to the determination of the lifetime of the device as shown in Fig. 5.

3.18 Lifetime of a Medical Device

The lifetime of a device is the useful life of a device, consisting of the duration of actual use before the device becomes incapable of achieving its intended use. In other words, its stability to withstand cleaning and sterilization during repeated use.

Testing should be done to estimate the shelf life of the product, which is the time at which the product naturally underperforms when newly manufactured, and its quality and accuracy cannot be assured when collecting the results.

4 Summary

From this chapter, we can extract that:

- Understanding the context of marketing a medical product highlights the importance of segmenting it into distinct phases.
- A medical product can cause great risks in a user's life by having a direct impact on their health.
- Following regulations and having the approval of a regulatory body is unquestionable for marketing this type of product.

When a researcher/engineer has the intention to follow this path, one must take into account the importance of all the topics described in this chapter, as these are the key words for the correct way to perform the technology transfer from research to industrialization. To help navigate this process, the checklist in Fig. 6 can serve as a guide, ensuring persistence through this long and bureaucratic journey, which is ultimately worthwhile.

Checklist: Technical Documentation

- ☐ Introduction:
  - ☐ Scope
- ☐ Administrative Information:
  - ☐ Manufacturer information
  - ☐ Date and version of the document
  - ☐ Applicable legislation
  - ☐ Medical device identification
  - ☐ Medical device classification
- ☐ Technical Documentation:
  - ☐ Medical device description
  - ☐ Intended use
  - ☐ State-of-the-art
  - ☐ Sales, complaints, and post-market surveillance
  - ☐ EU declaration
  - ☐ Technical standards and common specifications
  - ☐ General safety and performance requirements
  - ☐ Manufacturing process and subcontractors
  - ☐ Information for users
  - ☐ Design, validation, and verification
  - ☐ Risk management
  - ☐ Clinical evaluation
  - ☐ Clinical performance and safety summary
  - ☐ Post-market surveillance and post-market clinical follow-up
  - ☐ Biological safety
  - ☐ Packaging
  - ☐ Expiration and stability tests
  - ☐ Lifetime of medical device
- ☐ Annex List

Fig. 6 Checklist for the Technical Documentation of medical devices

Certification Glossary

ISO 13485 - Medical devices - Quality management systems - Requirements for regulatory purposes. 183

ISO 14971 - Medical devices - Application of risk management to medical devices. 183

ISO 24971 - Medical devices - Guidance on the application of ISO 14971. 183

Final Delivery - Date for the start of commercialization of the medical device. 177

Technical Documentation - Central part of the approval process for medical devices. Without this documentation, a manufacturer cannot carry out the conformity assessment procedure with their medical device, so the product cannot be approved. 175, 176, 178, 179, 187

Clinical Evaluation Report (CER) - Documents the conclusion of a clinical evaluation of your medical device. 175, 179, 183

European Medical Device Nomenclature (EMDN) - Nomenclature identifier based on key principles jointly established by the European Commission and EU regulators. 176, 179

Failure Mode and Effects Analysis (FMEA) - Systematic, proactive method of evaluating a process to identify where and how it is likely to fail and assessing the relative impact of different failures in order to identify the parts of the process most in need of change. 174

General Safety and Performance Requirements (GSPR) - States that the devices shall be designed and manufactured in such a way that the safety of patients and users shall not be compromised. 181

Medical Devices Regulation (MDR) - Aims to ensure the smooth functioning of the internal market as regards medical devices. 184

Post-market Surveillance (PMS) - Requirements that require manufacturers to adequately monitor their medical devices after they have been approved for sale. 180, 181, 184

Proof of Concept (PoC) - A Proof of Concept is meant to determine the feasibility of the idea or to verify that the idea will function as envisioned. 174–176

Regulation (EU) 2017/745 - EU regulation on clinical research and the sale of medical devices for human use. 175, 179, 181, 183

Risk Analysis Matrix (RAM) - Tool that helps visualize the probability versus the severity of a potential risk. 174

Risk Report (RR) - Method of identifying risks linked to or potentially impacting an organization's business processes. 174, 175, 179

Requirements Traceability Matrix (RTM) - Document that demonstrates the relationship between requirements and other artifacts. 175

Summary of Safety and Clinical Performance (SSCP) - Intended to provide public access to an up-to-date summary of clinical data and other information about the safety and clinical performance of the medical device. 184

Unique Device Identifier (UDI) The FDA established the unique device identification system to adequately identify medical devices sold in the United States from manufacturing through distribution to patient use. 176, 179

Acknowledgments This work was partially funded by FCT/MECI through national funds (FCT project references and DOI identifiers https://doi.org/10.54499/2022.12369.BD and https://doi.org/10.54499/2022.10245.BDANA) and when applicable co-funded EU funds under project UID/50008: Instituto de Telecomunicações. It was also co-supported by the European Regional Development Fund (FEDER), through the Lisbon Regional Programme (LISBOA 2030) of the Portugal 2030 framework—LISBOA2030-FEDER-01318700 (ComSense).

References

1. J.U. Ocampo, P.C. Kaminski, Medical device development, from technical design to integrated product development. J. Med. Eng. Technol. **43**(5), 287–304 (2019)
2. Z. Xu, Y. Dang, P. Munro, Y. Wang, A data-driven approach for constructing the component-failure mode matrix for FMEA. J. Intell. Manuf. **31**(1), 249–265 (2020)
3. K. Singh, P. Selvam, Medical device risk management, in *Trends in Development of Medical Devices* (2020), pp. 65–76
4. H.-C. Liu, Fmea for proactive healthcare risk analysis: a systematic literature review, in *Improved FMEA Methods for Proactive Healthcare Risk Analysis* (2019), pp. 15–45
5. Y. Al Khamisi, J. Hernandez, M. Khan, Developing a knowledge-based lean six sigma model to improve leadership's performance in healthcare environment (2018)
6. J.K. Aronson, C. Heneghan, R.E. Ferner, Medical devices: definition, classification, and regulatory implications. Drug Safety **43**(2), 83–93 (2020)
7. P.J. Ogrodnik, *Medical Device Design: Innovation from Concept to Market* (Academic Press, New York, 2019)
8. T. Yaqoob, H. Abbas, M. Atiquzzaman, Security vulnerabilities, attacks, countermeasures, and regulations of networked medical devices—a review. IEEE Commun. Surv. Tutorials **21**(4), 3723–3768 (2019)
9. Official Journal of the European Union, Regulation (EU) 2017/745 of the European Parliament and of the Council of 5 April 2017 (2017). https://eur-lex.europa.eu/legal-content/EN/TXT/PDF/?uri=CELEX:32017R0745
10. A. Badnjević, L. G. Pokvić, A. Deumić, L.S. Bećirović, Post-market surveillance of medical devices: a review. Technol. Health Care **30**(6), 1315–1329 (2022)
11. M. Manu, G. Anand, A review of medical device regulations in India, comparISOn with European union and way-ahead. Perspect. Clin. Res. **13**(1), 3 (2022)
12. A.G. Fraser, E.G. Butchart, P. Szymański, E.G. Caiani, S. Crosby, P. Kearney, F. Van de Werf, The need for transparency of clinical evidence for medical devices in Europe. The Lancet **392**(10146), 521–530 (2018)
13. B. Maestroni, General requirements for food safety analysis, in *Analytical Methods for Agricultural Contaminants* (Elsevier, Amsterdam, 2019), pp. 5–23
14. ISO, ISO 14971:2019—medical devices: application of risk management to medical devices (2017). https://www.ISO.org/standard/72704.html
15. ISO, ISO 13485—medical devices: quality management systems—requirements for regulatory purposes (2016). https://www.ISO.org/standard/59752.html.

16. ISO, ISO/TR 24971—medical devices—guidance on the application of ISO 14971 (2020). https://www.ISO.org/standard/74437.html
17. X. Liu, G. Sachdeva, H. Ibrahim, M. Charalambides, A.K. Denniston, Clinical evaluation of AI in medicine, in *Artificial Intelligence in Medicine* (Springer, 2022), pp. 645–660
18. ISO, ISO 10993-1—biological evaluation of medical devices—part 1: evaluation and testing within a risk management process (2018). https://www.ISO.org/standard/68936.html

Case Study: Towards Firefighter Monitoring in the Real World

Camila Abreu and Hugo Plácido da Silva

1 Introduction

Firefighters encounter a multitude of life-threatening challenges daily in their mission to protect lives and property, encompassing emotional, psychological, and environmental hazards. Some of the critical conditions they often face include:

- *Exposure to hazardous materials*, which heightens the likelihood of developing respiratory and chronic diseases
- *Physical strain and health risks*, as they often endure extreme heat with limited cooling options and smoke-filled environments, significantly increasing the risk of fatal cardiac events
- *Posttraumatic stress* due to the high-pressure nature of their work
- *Water-related incidents*, which can result in hypothermia
- *Fire-related incidents*, where the dangers of severe burns and toxic gas inhalation are ever-present

These challenges underscore the critical need for real-time health monitoring in firefighters [1]. However, studying these health risks in real-world scenarios is difficult due to the unpredictability and time-sensitive nature of emergency situations [2].

C. Abreu (✉) · H. P. da Silva
Instituto Superior Técnico, Universidade de Lisboa, Lisbon, Portugal

IT – Instituto de Telecomunicações, Lisbon, Portugal
e-mail: camila.f.abreu@tecnico.ulisboa.pt

H. P. da Silva et al. (eds.), *Open Source Biomedical Engineering*,
https://doi.org/10.1007/978-3-032-03655-1_10

In this chapter, we describe the SafeFire case study, in which one component is the development of a smart undergarment to ameliorate these needs, reducing the risks associated with firefighters' missions through vital signs monitoring.[1]

The SafeFire smart undergarment device builds upon the ScientISST CORE hardware base (Chap. 2), enabling the acquisition of a single-lead electrocardiogram (ECG) and either its recording to a memory card or the real-time streaming to the ScientISST SENSE API (chap. 3). For signal processing and feature extraction, the BioSPPy toolbox (Chap. 4) is used.

This chapter is divided into several sections. Section 2 provides an overview of existing wearable monitoring systems for firefighters. Section 3 outlines the methodology, beginning with an overview of the smart undergarment system (Sect. 3.1). This includes the various types of garments, the materials used for both the garments and electrodes, the interface connectors for electronics integration, and the encapsulation of the acquisition unit. Section 3.2 details the experimental protocol followed during testing at fire stations, while Sect. 3.3 focuses on ECG signal processing, including advanced filtering techniques using accelerometer data and an algorithm for assessing ECG signal quality. Section 4 presents the chapter's results, including the performance of the ECG quality assessment algorithm and an analysis of cardiac wave metrics across various activities, such as rest, arm movements, and running, captured using the smart undergarment. Finally, Sect. 5 provides a summary of the key insights, discusses the limitations, and outlines potential directions for future research.

2 Related Work

Firefighter monitoring using smart garments has been a topic of continued interest over the past years. In this section, we will provide a brief state-of-the-art review.

Research on physiological monitoring in firefighting mainly focuses on assessing the physical demands, cardiac metrics, and stress levels encountered by firefighters in real-world conditions [2]. Heart rate is a critical measure for evaluating physical activity and potential overload, making it a key indicator in this context. Recent advancements have introduced additional metrics, such as respiratory rate and core body temperature, which provide a more comprehensive understanding of a firefighter's physiological state during operations. These enhancements improve the detection of physical fatigue, leading to more effective management of work-rest cycles and enhanced overall safety [3].

The integration of smart garments into firefighting gear is subject to stringent regulations. For instance, according to the European norm EN ISO 11612,[2] textile materials must not undergo changes when exposed to 180°C for 5 minutes [4].

[1] https://www.scientisst.com/projects/safefire

[2] https://marinatextil.com/pt/norma-tecido/iso-11612-norma-tecido-de-protecao

Additionally, firefighters' undergarments must contain less than 2% elastane to comply with safety standards.

Vital Jacket Vital Responder[3] was a highly prolific project, focused on monitoring stress, vital signs, and fatigue in First Responders (FR). Initially, it used a seamless elastic shirt with embedded sensors and a separate jacket for electronics. The sports version later evolved into a simple t-shirt with integrated electronics. The Vital Jacket features a shirt embedded with an electronic device, including Electrocardiogram ECG sensors for heart activity and a 3-axis accelerometer for monitoring physical activity and posture [5].

The Vital Jacket uses 1, 3, or 5 ECG leads to measure signals, transmitting data in real time to devices or cardiology systems via wireless networks, while also storing data on an SD card for offline analysis. However, adapting this wearable technology to adverse conditions can lead to low-quality ECG readings due to damaged electrodes, hardware malfunctions, and/or incorrect timestamp annotations [6].

Another important consideration is that the fabric of the Vital Jacket originally contained 28% elastane, which was later reduced to 2%. While a higher elastane content is beneficial for achieving the desired fit, it does not comply with the regulations for firefighters' undergarments.

In 2007, the technology was licensed to Biodevices S.A., achieving ISO9001 and ISO13485 certifications and approval as a medical "Ambulatory ECG device" under the European MDD directive. It is the first wearable cardiology long-term monitor to earn this certification globally [5].

Equivital The Equivital project focuses on mobile human monitoring for firefighters and soldiers. It emphasizes real-time monitoring to prevent heat-related injuries that impair cognitive function and can be life-threatening. The system provides comprehensive visibility into physiological and psychological states, enabling timely decisions by commanders and medics.[4]

Equivital products, such as the eq02+ LifeMonitor and eq LifeBand, use lightweight, comfortable fabric sensors integrated into clothing or harnesses.

The eq02+ LifeMonitor[5] offers a range of wired and wireless sensors for multiparameter monitoring, including heart rate, galvanic skin response, respiratory rate, skin temperature, and movement. It features an ECG, respiratory monitor, medical-grade thermometer, and tri-axial accelerometer. With high data quality, minimal loss, and up to 48 hours of battery life, it supports up to 50 days of continuous logging. It is FDA 510K approved, CE marked, waterproof (IP67), and designed for extreme conditions.

---

[3] http://www-archive.inesctec.pt/cber/research/bioinstrumentacao/vital-responder/

[4] https://equivital.com/mobile-human-monitoring

[5] https://equivital.com/products/eq02-lifemonitor

The eq LifeBand[6] monitors body temperature to prevent heat stress, providing real-time or post-event alerts for hydration and customized work-rest cycles based on environmental conditions. It complies with ACGIH heat stress guidelines.

ActionSAFE The ActionSAFE project, led by the Instituto de Ciência e Inovação em Engenharia Mecânica e Engenharia Industrial (INEGI) at the University of Porto, developed wearable sensors integrated into firefighter uniforms to monitor physiological health. The system tracks internal temperature, heart rate, physical activity, and respiratory rate to assess fatigue levels in real time during firefighting operations.[7] Data is processed through an online platform, accessible via computer or smartphone, with alarms that alert when a firefighter reaches extreme physical exhaustion.[8]

3 Methodology

The SafeFire project tackles the challenges of monitoring firefighters by selecting suitable materials and developing effective data processing methods. A key focus is finding undergarment and electrode materials that meet EU safety regulations—less than 2% elastane and high-temperature resistance—while maintaining accurate data acquisition. The fabric must be elastic, snug, and heat-resistant and allow free movement. To address these needs, SafeFire evaluates material options and creates algorithms to filter noisy ECG data.

3.1 Smart Undergarment

Overview The smart undergarment is part of a comprehensive data acquisition system, illustrated in Fig. 1. This undergarment includes embedded electrodes that capture the user's cardiac electrical signals. These signals are processed by an ECG sensor and transmitted to the ScientISST device, which either sends the data via Bluetooth or stores it on an SD card, depending on the configuration at startup. When transmitted via Bluetooth, the data can be visualized in real time through the ScientISST SENSE Web application and saved as a .csv file for later analysis.

Types of Garments To achieve the best fit for reliable physiological signal measurement during various activities, three different types of garments were evaluated:

---

6 https://equivital.com/products/eq-lifeband

7 https://www.enb.pt/publicacao.php?id=649

8 https://www.reconquista.pt/articles/actionsafe-prototipo-em-teste-para-monitorizar-saude-dos-bombeiros-

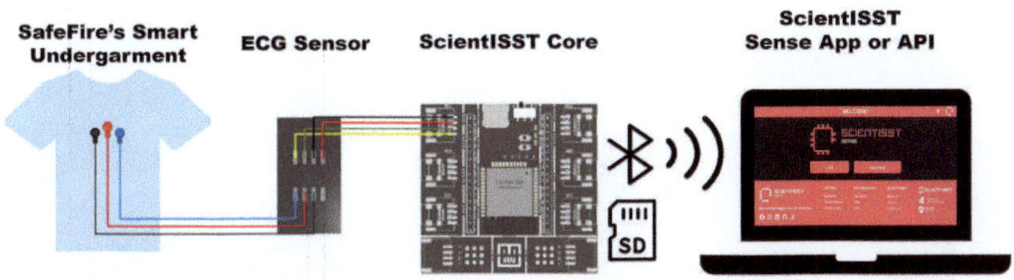

Fig. 1 Schematic illustration of the acquisition system integrated into SafeFire's smart undergarment. Adapted from [4]

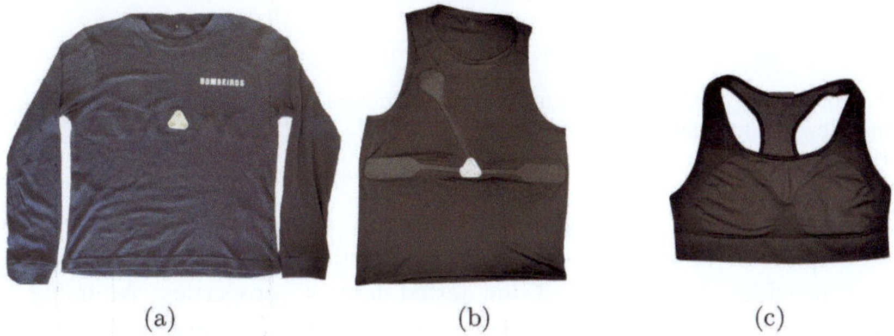

Fig. 2 Different undergarment prototypes evaluated for signal acquisition. (**a**) Aramid undergarment. (**b**) Singlet. (**c**) Sports bra/vest

- **Aramid Undergarment**: This shirt, made from aramid, is resistant to high temperatures and provides good mobility. However, it does not fit tightly enough during medium-/high-intensity activities, which can lead to signal noise (Fig. 2a).
- **Singlet**: This sleeveless garment reduces noise associated with electrode movement when the arms are raised. However, it still lacks the necessary tightness to ensure consistent signal quality during intense activities (Fig. 2b).
- **Sports Bra/Vest**: This garment offers the desired snug fit, eliminating noise from arm movements and reliably detecting heartbeats even during high-intensity activities such as running (Fig. 2c).

Materials of Garments and Electrodes The choice of material is critical for ensuring comfort and safety in high-temperature environments. The prototype smart undergarment is crafted from meta-aramid, a material extensively used in industries like metallurgy, construction, and fire protection. This makes it ideal for specialized protective clothing.[9] Meta-aramid is renowned for its flame resistance, non-melting and non-dripping properties, and inability to support combustion. These features, combined with its comfort, make it well-suited for firefighters (Fig. 3).

[9] https://www.tchaintech.com/CDETAIL/case-firefighter-aramid-fabric

(a) (b)

Fig. 3 Prototype of SafeFire's smart undergarment featuring metal-plated fabric electrodes. (**a**) External view. (**b**) Internal view.

In addition to evaluating garment materials, various electrode fabrics were assessed for their effectiveness in signal acquisition during dynamic conditions. Table 1 provides a summary of the tested fabrics' properties, including their mechanical and electrical characteristics and their high-temperature resistance, as evaluated by the Associação para o Desenvolvimento da Aerodinâmica Industrial (ADAI). Although Lycra was initially considered, it was excluded from the study due to its failure to withstand temperatures up to 200 °C.

Interface Connectors for Electronic Integration The interface between the textile and the electronic components is crucial for the reliable operation of the smart undergarment. The connectors, produced by laser cutting beforehand, are heat-pressed onto the garment at specific locations, with silver paste applied to enhance conductivity. The 3D-printed ScientISST connector (Fig. 4a) is then sewn onto the garment, establishing a connection with the ScientISST device via cables (Fig. 4b).

Acquisition Unit The ScientISST device [7] is compact and designed to be carried in a small pocket, as shown in Fig. 5. Figure 5c shows the opening for removing and replacing the SD card, as well as the connection for charging the device's battery. By removing the SD card and powering on the board, real-time data can be collected via Bluetooth. Inserting the SD card and subsequently powering on the board allows for data storage on the SD card for later analysis. The lower right opening in Fig. 5b is designed for connecting the cables (Fig. 4b) that run from the smart undergarment to the ScientISST device. Additionally, the open state of the encapsulated ScientISST device is displayed in Fig. 6, offering an internal view of the board and battery configuration.

Table 1 Properties of Digikey Ripstop[a], Sigracell GFD 2.5[b], Sigracell GFD 4.65[b], and Raypad fabrics[c]

Fabric	Mechanical properties	Electrical properties	Temperature resistance	Water resistance	Characterization
Digikey Ripstop	High tensile strength, low elongation	Excellent conductivity; sheet resistance $< 0.02\Omega/\square$[d]	Moderate (up to 380 °C)	High (water repellent)	Durable and breathable, used in outdoor gear
Sigracell GFD 2.5	High flexibility, good strength	Good conductivity; sheet resistance $<0.1\Omega/\square$[d]	High (up to 680 °C)	Moderate	Microporous carbon electrode, used in sensors
Sigracell GFD 4.65	Moderate flexibility, more robust	Good conductivity; sheet resistance $<0.15\Omega/\square$[d]	High (up to 680 °C)	Moderate	More robust microporous carbon electrode with enhanced thickness
Raypad	Moderate elasticity, wear-resistant	Excellent conductivity; sheet resistance $<0.05\Omega/\square$[d]	Moderate (up to 270 °C)	High (water-resistant)	Versatile, used in electronics and sportswear

[a] https://resources.kitronik.co.uk/pdf/2716-conductive-fabric-riptop.pdf
[b] https://www.sglcarbon.com/data/pdf/SGL-Datasheet-SIGRACELL-Battery-Felts-EN.pdf
[c] https://lessemf.com/product/raypad/
[d] The sheet resistance values were measured in laboratory tests conducted by the Laboratório de Estudos sobre Incêndios Florestais in Portugal

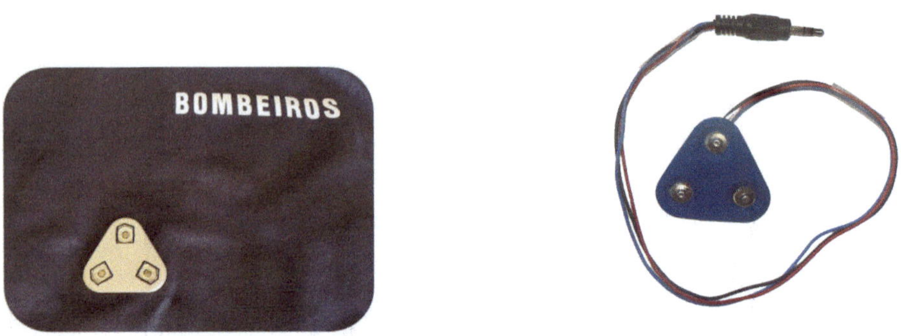

(a) (b)

Fig. 4 Components of the connection interface for SafeFire's smart undergarment. (**a**) Connector integrated into SafeFire's smart undergarment. (**b**) Cables used with the SafeFire system

(a) (b) (c)

Fig. 5 Overview of the ScientISST device encapsulation process. (**a**) Design of the two components used for encapsulation. (**b**) Completed encapsulation of the ScientISST device using the two components. (**c**) Opening for SD card access and battery charging

Fig. 6 Internal view of the ScientISST device showing the ScientISST board and battery arrangement

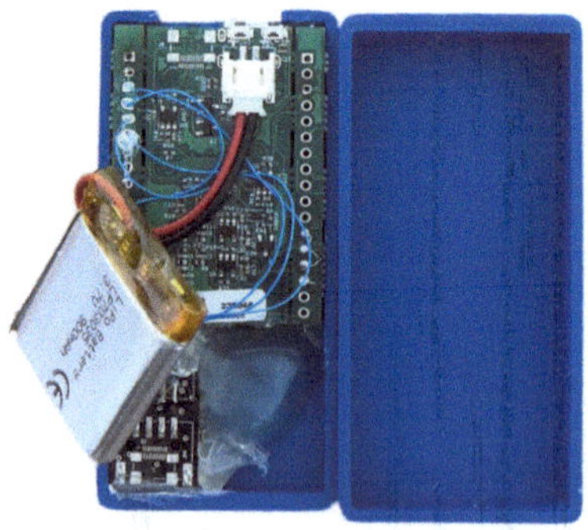

3.2 Testing Protocol

To test the different variables (type of undergarment, electrode fabric), tests were conducted at three fire stations and in a field setting where all the firefighters who participated in the station tests gathered.

Initially, the following protocol was carried out at each fire station while the firefighters wore the smart undergarment:

1. Rest for 1 minute
2. Arm movements:
   (a) 180° rotation of the right and left arms, alternately, in the frontal plane
   (b) 180° rotation of the right and left arms, alternately, in the sagittal plane
   (c) 120° trunk flexion
   (d) 45° trunk hyperextension
3. Running for 1 minute

For the fire station tests, it was necessary to observe the signal in real time, and due to the proximity to the firefighters, data collection was done using the ScientISST API via Bluetooth for data transmission.

In the field tests, the procedure was simplified to evaluate the system's performance under real conditions. The ECG was recorded during 5 minutes of rest before and after the field trials, which involved fire extinguishing tasks and during the field trials themselves. For the rest periods before and after the trials, data was collected using the ScientISST API, while during the field trials, data was stored on an SD card.

3.3 ECG Signal Processing

Challenges in ECG Monitoring ECG signals from wearable devices are typically noisier than those captured in clinical settings, due to factors such as electrode impedance, movement artifacts, and breathing interference. As a result, it is crucial to develop filtering and quality assessment algorithms for these signals.

High-Intensity Activity Filtering One approach to filtering involves excluding periods of high-intensity activity, and it consists of the following steps:

1. **Accelerometer data integration**: Using tri-axial accelerometer data in addition to the ECG sensor data, which are already integrated into the ScientISST device.
2. **Variance calculation**: Determining activity levels by calculating the variance of the accelerometer's magnitude.
3. **Threshold application**: Samples with accelerometer magnitude variance exceeding a specified threshold are classified as high-intensity periods, and the corresponding ECG data for these periods are excluded from analysis.

Figure 7 provides an example of simultaneous ECG signal acquisition with tri-axial accelerometer data using the ScientISST device, demonstrating the application of the described filtering method.

ECG Quality Assessment Accurate ECG quality assessment is crucial for ensuring reliable data interpretation and effective decision-making. To address this, the

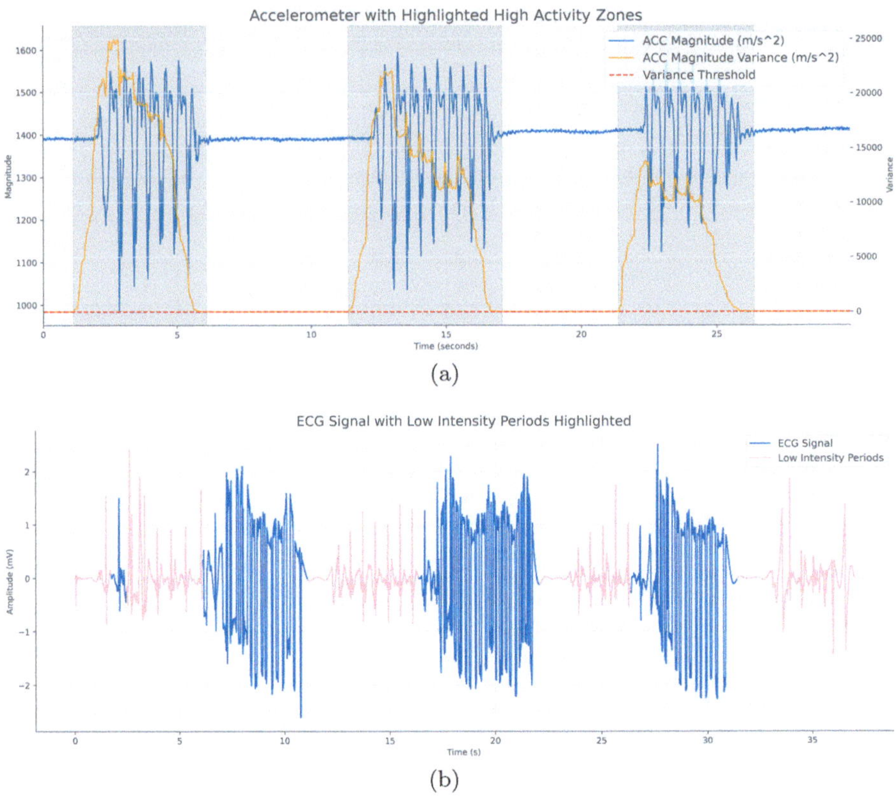

Fig. 7 Example of filtering using accelerometer data recorded simultaneously with ECG data using the ScientISST Cardio at 100 Hz: (**a**) shows the accelerometer data with high activity periods highlighted, and (**b**) shows the ECG signal with low-intensity periods highlighted

quality of ECG signals is evaluated using an assessment algorithm trained on the ScientISST MOVE dataset [8]. This dataset is notable for its extensive and diverse collection of wearable biosignal recordings, including a broad range of annotated activities that simulate real-life conditions (e.g., lifting a chair, walking, running). The ECG sensor utilized in the dataset was attached to a Polar chest band with dry conductive textile contacts, replicating the conditions encountered with SafeFire's smart undergarment.

The quality assessment is performed by a supervised machine learning algorithm that leverages eight features, detailed in Table 2, and incorporates Signal Quality Indexes (SQIs) from the BioSPPy library (Chap. 4) [9].

Among these, the ZZ2018 function from the `ecg.py` module of BioSPPy plays a central role. This function classifies ECG signal quality into three categories: "Excellent," "Barely Acceptable," and "Unacceptable." This classification is based on four specific SQIs (*qSQI*, *pSQI*, *kSQI*, and *basSQI*) available in the ecg.py module of BioSPPy. Additionally, the Quality.py module of BioSPPy includes

Table 2 Extracted ECG signal features for model training, divided into two domains: time and SQIs

Domain	Feature	Explanation
Time	Standard deviation of RR intervals (SDRR)	Standard deviation of the intervals between consecutive R-peaks
	Variability of R-peaks compared to maximum amplitude (MaxRR)	Ratio of the maximum amplitude of R-peaks to the maximum amplitude of the signal
	Inter-beat variability (InterV)	Average standard deviation across individual heartbeats
	Intra-beat variability (IntraV)	Average standard deviation of the template beats
	Flatline percentage (FlatP)	Percentage of the signal where the amplitude is below a certain threshold
SQI	Quality	Numeric representation of overall ECG quality based on multiple SQIs (*qSQI*, *pSQI*, *kSQI*, and *basSQI*)
	hosSQI	Combines the absolute value of the skewness quality index (*sSQI*) and the kurtosis quality index (*kSQI*)
	cSQI	Ratio of the standard deviation to the mean of RR intervals

further SQIs, such as *cSQI* and *hosSQI* [10, 11], which are also incorporated into the algorithm.

Following the evaluation of these metrics for each ECG segment, the segment is categorized as either "Acceptable" or "Unacceptable."

4 Results

ECG Quality Assessment
By utilizing the eight features (Table 2), we achieved 88% accuracy and an 87% F1 score in classifying a 10-second ECG segment as acceptable or unacceptable.

Real World Testing
Analyzing the results from tests conducted at the fire stations, we observed different levels of variability in metrics that evaluate the intervals and amplitudes of cardiac waves throughout the acquisition period. These results are represented in a grouped boxplot for three different scenarios: rest, arm movement, and running (Fig. 8).

At rest, the variability in both wave amplitude and interval metrics is minimal compared to the other activities. This stability is particularly clear in the PR, QRS, and QT intervals, as well as the amplitudes of the P, S, and T waves.

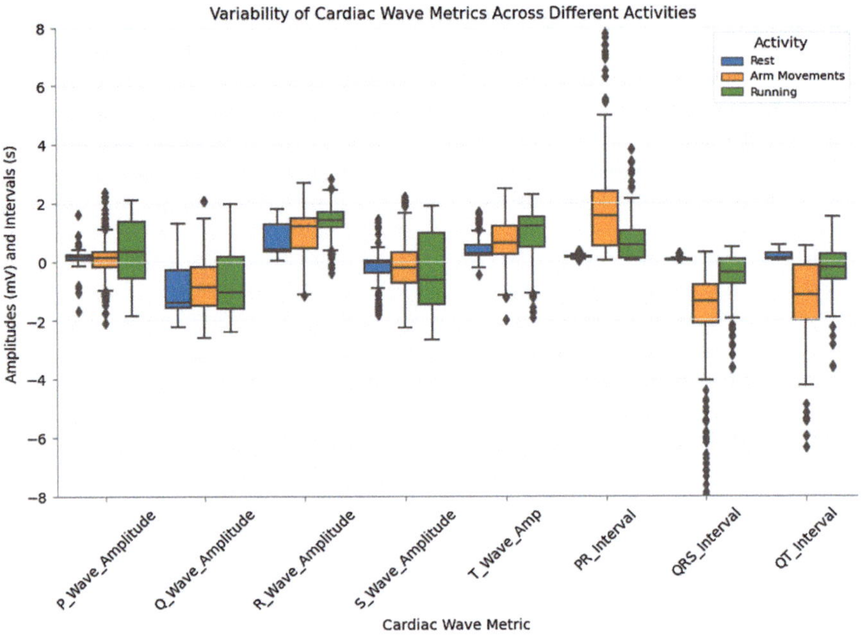

Fig. 8 Grouped boxplot illustrating the variability of cardiac wave metrics (P, Q, R, S, and T wave amplitudes in millivolts [mV] and PR, QRS, and QT intervals in seconds [s]) during ECG acquisition across different activities (rest, arm movements, running) for firefighters. The y-axis represents the metric values in mV or seconds, depending on whether it corresponds to an amplitude or interval, respectively, while the x-axis lists the different cardiac wave metrics

During arm movements, the variability in wave amplitude metrics increases, though it remains less pronounced than during running. Interval metrics show even greater variability compared to running, likely due to movement-induced noise.

Running causes the highest variability in wave amplitude metrics, especially in the R and T waves, which show numerous outliers. Although arm movements have more outliers in interval metrics, running also introduces significant variability.

In summary, rest results in the most stable cardiac metrics, as expected. Arm movements introduce moderate variability, mainly due to motion artifacts. Running leads to the highest variability, reflecting the heart's strong response to intense exercise and the noise caused by electrode movement along the skin, as well as occasional loss of skin contact.

Overall, these results show how our prototype can be used for real-time monitoring and assessment of cardiac health in firefighters, providing valuable insights into the impact of different physical activities on heart function.

5 Summary

After reading this chapter, it is clear that the use of smart undergarments for monitoring firefighters holds significant promise. These undergarments are noninvasive, comfortable, and "invisible," combining a compact size with high performance. The process requires careful selection of the undergarment type and electrode materials, along with precise ECG data signal processing. The key points covered in this chapter are summarized below:

- The SafeFire project aims to enhance firefighter safety by providing continuous real-time monitoring of vital signs using a smart undergarment equipped with dry textile electrodes.
- The system integrates with the ScientISST platform, allowing ECG data acquisition both online via Bluetooth and offline through an SD card, facilitating immediate risk assessment and post-mission health evaluations.
- The smart undergarment is made of aramid, a flame-resistant material, and features built-in electrodes to detect cardiac signals, which are then processed and transmitted by the ScientISST device.
- The project involved evaluating different types of undergarments and electrode materials and developing algorithms to filter out noisy ECG segments caused by movement and other factors.
- Based on the BioSPPy toolbox, advanced filtering techniques using accelerometer data and machine learning algorithms were employed to classify ECG signal quality and exclude high-intensity activity periods.
- The system achieved 88% accuracy and 87% F1-score in classifying ECG segments as acceptable or unacceptable, demonstrating its potential for effective real-time health monitoring in firefighters.
- Tests conducted at fire stations and in field conditions showed varying levels of ECG signal stability, with rest periods providing the most stable metrics and running causing the highest variability due to intense physical activity and electrode movement.

However, there are limitations in monitoring firefighters' vital signs, resulting in a lack of datasets for ECGs of firefighters on duty. These obstacles include high temperatures and radiation, which can cause sensor errors, and moisture, which reduces sensor lifespan and complicates long-term monitoring [12]. Additionally, the harsh conditions of firefighting introduce noise from body movement, breathing, and environmental factors [13].

Future research will prioritize the integration of multiple sensors to comprehensively monitor the diverse health impacts associated with firefighting. Given that firefighting activities lead to significant increases in heart rate, core temperature, blood lactate levels, and psychological stress, real-time psychophysiological monitoring is essential. This approach will enable the early detection of chronic stress by combining both psychological and physiological data, allowing for more accurate and effective stress monitoring in natural, high-stress environments like firefighting [2].

To enhance firefighter safety, upcoming developments include a smart undergarment designed to monitor vital signs in both online and offline modes. In online mode, real-time data transmission via Bluetooth will reduce mission risks and enable immediate emergency interventions. Offline mode will store data on an SD card for post-mission analysis, helping to assess health conditions and refine machine learning algorithms for improved safety outcomes.

Acknowledgments This work was partially funded by FCT/MECI through national funds (FCT projects PCIF/SSO/0163/2019 SAFEFIRE with DOI: https://doi.org/10.54499/PCIF/SSO/0163/2019 and UIDB/50008/2020 with DOI: https://doi.org/10.54499/UIDB/50008/2020) and when applicable co-funded EU funds under project UID/50008: Instituto de Telecomunicações. It was also co-supported by the European Regional Development Fund (FEDER), through the Lisbon Regional Programme (LISBOA 2030) of the Portugal 2030 framework - LISBOA2030-FEDER-01318700 (ComSense).

References

1. J. Li, C. Brown, D.J. Dzikowicz, M.G. Carey, W.C. Tam, M.X. Huang, Towards real-time heart health monitoring in firefighting using convolutional neural networks. Fire Safety J. **140**, 103852 (2023)
2. D. Bustos, J.C. Guedes, J. Santos Baptista, M. Vaz, J. Torres Costa, R.J. Fernandes, in *Physiological Monitoring Systems for Firefighters (A Short Review)*. (Springer, Cham, 2022), pp. 293–305
3. D. Bustos, R. Cardoso, D.D. Carvalho, J. Guedes, M. Vaz, J. Torres Costa, J. Santos Baptista, R.J. Fernandes, Exploring the applicability of physiological monitoring to manage physical fatigue in firefighters. Sensors **23**(11), 5127 (2023)
4. E.R. Varela, Smart Undergarment for Firefighter Monitoring under Dynamic Conditions. Master's Thesis, Instituto Superior Técnico, 2022
5. J.P.S. Cunha, B. Cunha, A.S. Pereira, W. Xavier, N. Ferreira, L. Meireles, Vital-Jacket®: a wearable wireless vital signs monitor for patients' mobility in cardiology and sports, in *2010 4th International Conference on Pervasive Computing Technologies for Healthcare* (2010), pp. 1–2
6. M. Coimbra, J.P. Silva Cunha, Vital Responder – wearable sensing challenges in uncontrolled critical environments, in *Sensor Systems and Software*, ed. by F. Martins, L. Lopes, H. Paulino (Springer, Berlin, Heidelberg, 2012), pp. 45–62
7. L. Pereira, F. de Melo, F.A. Santos, A.F. Ferreira, H.P. da Silva, *ScientISST CORE: A Novel Hardware Development Platform for Biomedical Engineering*, vol. 21 (HardwareX, 2025), p. e00630
8. J. Areias Saraiva, M. Abreu, A.S. Carmo, H. Plácido da Silva, A. Fred, ScientISST MOVE: annotated wearable multimodal biosignals recorded during everyday life activities in naturalistic environments. PhysioNet (2024)
9. P. Bota, R. Silva, C. Carreiras, A. Fred, H.P. da Silva, BioSPPy: a python toolbox for physiological signal processing. SoftwareX **26**, 101712 (2024)
10. Z. Zhao, Y. Zhang, SQI quality evaluation mechanism of single-lead ECG signal based on simple heuristic fusion and fuzzy comprehensive evaluations. Front. Physiol. **9**, 727 (2018)

11. M. Nardelli, A. Lanata, G. Valenza, M. Felici, P. Baragli, E. Scilingo, A tool for the real-time evaluation of ECG signal quality and activity: application to submaximal treadmill test in horses. Biomed. Signal Process. Control **56**, 101666 (2020)
12. F. Shakeriaski, M. Ghodrat, Challenges and limitation of wearable sensors used in firefighters' protective clothing. J. Fire Sci. **40**(3), 214–245 (2022)
13. Z. Alimbayeva, C. Alimbayev, K. Ozhikenov, N. Bayanbay, A. Ozhikenova, Wearable ECG device and machine learning for heart monitoring. Sensors **24**(13), 4201 (2024)

Case Study: Sympathia Sense

Afonso Fortes Ferreira, Helena Alves, and Hugo Plácido da Silva

1 Introduction

Wearable devices employing Electrodermal Activity (EDA) sensors offer valuable insights into the variations of the Autonomic Nervous System (ANS), making them useful in fields such as affective computing and health monitoring [1, 2]. The EDA signal in particular, is of interest due to its ability to reflect emotional and cognitive states through responses mediated by the Sympathetic Nervous System (SNS), a subsystem of the ANS [1]. While most wearables typically use EDA sensors at the wrist for its convenience (i.e., smartwatches contact the skin at the wrist when worn), other sites, such as the palms and the soles, can provide more responsive EDA signals due to the relatively higher density of sweat glands (and their respective efferent nerves) within these bodily locations [3]. Traditional wearables like wristbands or smartwatches can be perceived as intrusive by some users and, in some cases, stigmatizing. This can happen in some groups of users, such as patients with epilepsy or mental health disorders, potentially leading to low user acceptance or long-term adoption of such technologies [4]. It is, therefore, crucial to explore form factors that consider both sensor positioning and its impact on EDA signals, as well as usability aspects such as discreetness and comfort, which influence the acceptability of the device from the user's perspective [5].

A. F. Ferreira (✉) · H. Alves
Instituto Superior Técnico, Universidade de Lisboa, Lisbon, Portugal

INESC-MN - Instituto de Engenharia de Sistemas e Computadores para os Microsistemas e Nanotecnologias, Lisbon, Portugal
e-mail: afonsofortes@tecnico.ulisboa.pt

H. P. da Silva
Instituto Superior Técnico, Universidade de Lisboa, Lisbon, Portugal

IT - Instituto de Telecomunicações, Lisbon, Portugal

H. P. da Silva et al. (eds.), *Open Source Biomedical Engineering*,
https://doi.org/10.1007/978-3-032-03655-1_11

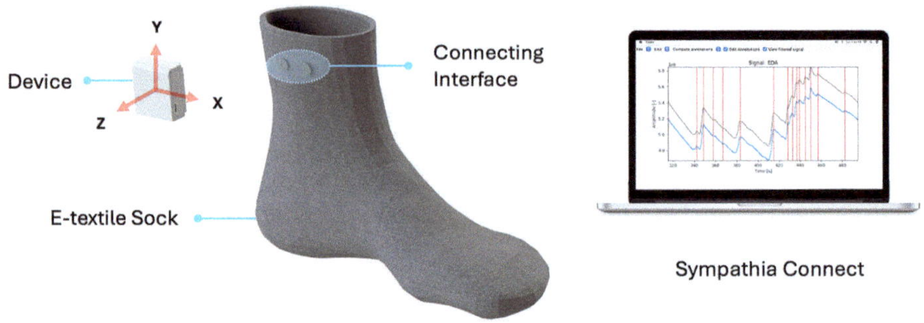

Fig. 1 Sympathia Sense. The e-textile sock is displayed on the left, featuring its connecting interface and detachable device. The overlaid 3-axis diagram represents the axes of the integrated acceleration sensor; this coordinate system corresponds to the joint coordinate system for the ankle, as recommended by the International Society of Biomechanics [6]. The Sympathia Connect application is shown on the right. This platform is a computer-based tool for biosignal acquisition and processing

Sympathia Sense provides a more socially acceptable and effective solution for EDA and body motion monitoring. As shown in Fig. 1, this system comprises a smart sock and a computer app, Sympathia Connect. The smart sock consists of the e-textile sock and the acquisition device, which is connected to the former through a dedicated e-textile interface. This interface establishes an electromechanical connection that is necessary to collect the EDA signals through the sock. Moreover, the device also uses an integrated 3-axial accelerometer to monitor body movement. The 3 axes of the acceleration sensor are shown in Fig. 1.

2 E-Textile Sock

2.1 Pattern Design and Optimization

The e-textile sock is manufactured by embroidering conductive yarn on a regular sock, similarly to the procedure described in a previous work [7]. The first step is to design the pattern for the conductive traces, which can be done in the sketch software Inkscape, as shown in Fig. 2a. The conductive traces were drawn by creating a spline with a width of 2 mm and adjusting the nodes to create a wavy pattern (in Inkscape we used the Pen and Node tools). This wavy geometry allows the sock to extend without stiffening its mesh, an important requirement for socks. The remaining parts of the sketch consist of the "connecting pads" and the electrodes. The connecting pads are circles tailored to match the footprint of the connectors, so these can be inserted through the fabric, whereas the electrodes consist of two 10 × 15 mm rectangles separated by 5 mm. The design can be exported in SVG or PDF and opened in a second software, sometimes specific to the embroidery machine, where the design is further refined according to specific embroidery parameters. Some

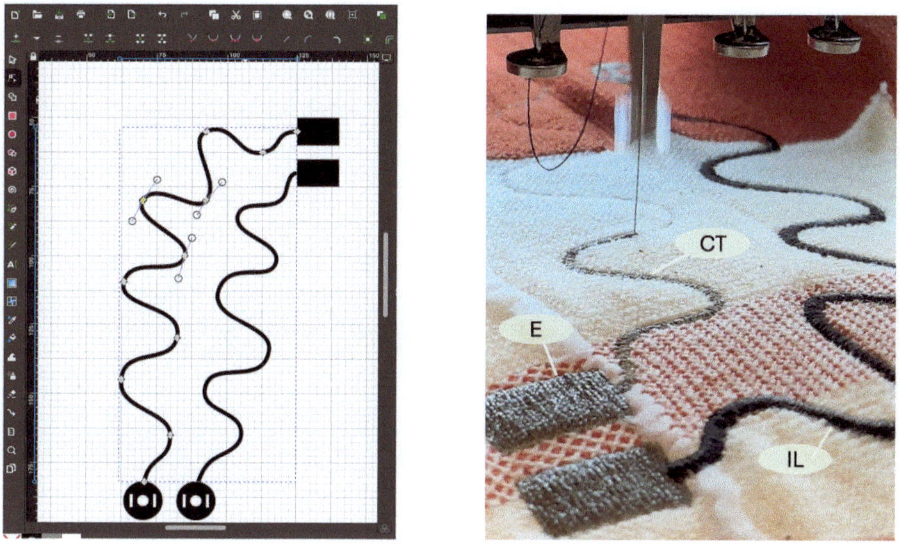

(a) (b)

Fig. 2 **(a)** Design software Inkscape. **(b)** Sock being embroidered. The figure provides an intermediate snapshot of the embroidery process. At the moment the image was captured, the trace on the left had yet to be embroidered with the insulating layer, highlighting the technique in progress. The key components are labelled similarly as in Fig. 3: conductive trace (CT), electrodes (E), and insulating layer (IL)

of these parameters include stitch type (satin, fill, and running stitch) and density (number of stitches per millimeter).

Most likely, embroidering e-textiles will require optimizing the design. In our case, this optimization involved the trade-off between the electrical resistance and the stretchability of the fabric. For instance, increasing the stitch density and the trace width leads to a decrease in resistance, but also increases fabric stiffness. Similarly, reducing the amplitude of the wavy pattern (i.e., by reducing the arc angle and the the radius of the arc) reduces the overall path of the conductive traces, decreasing its resistance, but also increasing stiffness. Beyond our specific example, optimizing e-textile embroidery requires testing on a case-by-case basis, as it is highly dependent on each design, embroidery software, type of conductive yarn, and embroidery machine.

2.2 Practical Implementation

Figure 2b shows the sock being embroidered. The sock is preferably compressive, to ensure conformability to the shape of the foot and, consequently, contact between

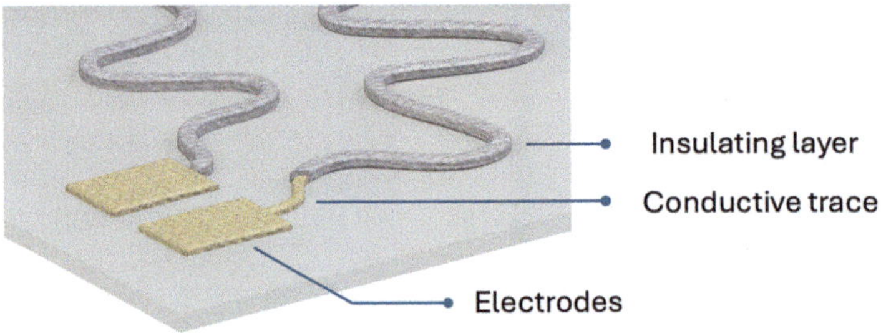

Fig. 3 Detailed view of the electrodes. The various layers of the sock are shown: base fabric (bottom, in gray), conductive trace and electrodes (in beige), and insulating layer (top, in gray)

the skin and the electrodes. It consists of a regular commercially available sock produced by circular knitting (weft knit mesh) with high elastomer content (e.g., Nylon), ensuring elasticity. For the conductive yarn, we used HC 40 (from Madeira Garnfabrik Rudolf Schmidt KG, Germany) which has an electrical resistance $<300\,\Omega$/m and is resistant to machine washing at 30 °C (89 °F). The sock is prepared by first bisecting it in the transverse plane to lay it flat. The resulting fabric is affixed to a stabilizer at the embroidery hoop using temporary adhesive, and the designed pattern is then embroidered. Figure 3 shows a close-up of the pattern at the inside of the sock. The bottom layer (golden fabric) consists of two conductive traces that extend in a wavy pattern and then form the two exposed rectangles, serving as EDA electrodes. This layer serves as the electrical connection to the device and establishes the interface between the sensor and the skin. The top layer consists of a pattern embroidered on top of the conductive traces, made of regular nonconductive yarn. This layer provides electrical insulation between the conductive traces and the skin, ensuring that the EDA signal is obtained from the foot location where the electrodes are placed.

3 Detachable E-Textile Connector

For the interface between the e-textile sock and the device, our approach uses magnetic Snap Buttons (SBs) to establish a detachable connection mechanism, which is shown in Fig. 4. Through this detachable mechanism the device is attached to the sock upon usage, or detached from it when the user needs to charge the device and/or wash the sock, as shown in Fig. 1. This mechanism is preferred over permanently fixing the device to the sock (e.g., sealing it with silicone) for its cost-effectiveness and sustainability [8]. Firstly, detachable connectors decouple the lifespan of the electronic unit from that of the e-textile sock. This is especially important in our form factor because socks are washed quite frequently (after each single use), when compared to other garments (e.g., pants and sweatshirts may not need to be washed every day). Secondly, specifically, magnetic SBs are inexpensive,

Fig. 4 Magnetic SBs in **(a)** open view and **(b)** section view (the magnetic force exerted by the female part on the male stud is illustrated by the pink arrows)

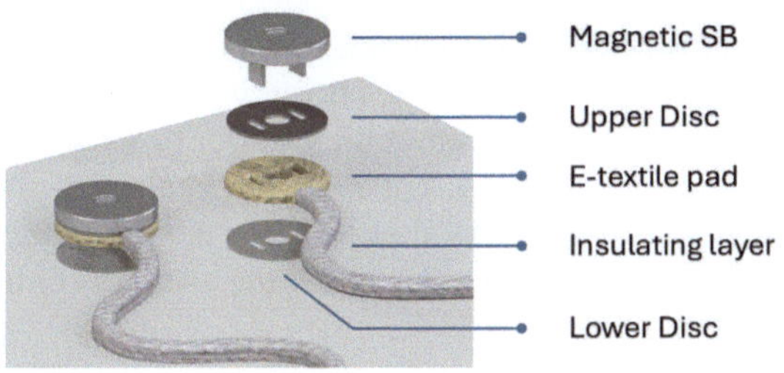

Fig. 5 Magnetic SBs integrated into the e-textile sock. The exploded view on the right highlights the assembly of the connectors, demonstrating how each component is secured to the fabric to ensure a reliable mechanical and electrical connection

widely available, and straightforward to implement both in the e-textile and the device.

On a practical level, the magnetic SBs are implemented both in the sock and at the device's enclosure. As shown in Fig. 5, at the sock, the female parts pierce through the fabric at the exposed e-textile pads of the pattern, ensuring physical contact and electrical conductivity with the pattern. To further ensure good conductivity, two metallic discs are sandwiched between the magnetic SBs and the sock's fabric, one on each side, after which the magnetic SBs are bent and locked in place. At the device, the respective SBs counterparts are crimped to the device's enclosure and electrically connected from the inside. When mated, the SBs on the sock exert and maintain a strong magnetic force on the SBs of the device, which is further secured by their matching protrusion/indentation (see the cross-section image in Fig. 4b). This attachment mechanism in magnetic SBs has demonstrated excellent electrical and mechanical stability even when under considerable loads and/or upon intense human body movement, preserving signal integrity [9].

4 Device

4.1 Computer-Aided Design

Figure 6a shows the device on its two sides: the side facing the sock (left) and the opposite side (right) which remains visible to the user after attaching it. Figure 6b shows the detailed assembly of the main unit in its exploded view. As shown, the device has a two-part enclosure given by the lid and the box. This enclosure houses the Printed Circuit Board (PCB), the battery, and the magnetic SBs, which are inserted halfway into the box.

The enclosure and the PCB are secured using screws and screw inserts, a method chosen over permanent fixation techniques to facilitate repairability, modification, and reusability. For 3D-printed enclosures made with PLA, the screw inserts can be placed by pressing them with a soldering iron (at a temperature >220 °C). The PCB is fixed to the box through 3 screws, preventing the PCB from moving inside it when connecting the USB-C cable from the exterior or when switching the on/off button. This fixation is shown in Fig. 7. As for the enclosure, the box and the lid use 4 screws to fix these two parts. As shown in Fig. 8, the screws are inserted from the side of the lid, making them unnoticeable to the user (as confirmed by the right image in Fig. 6a.

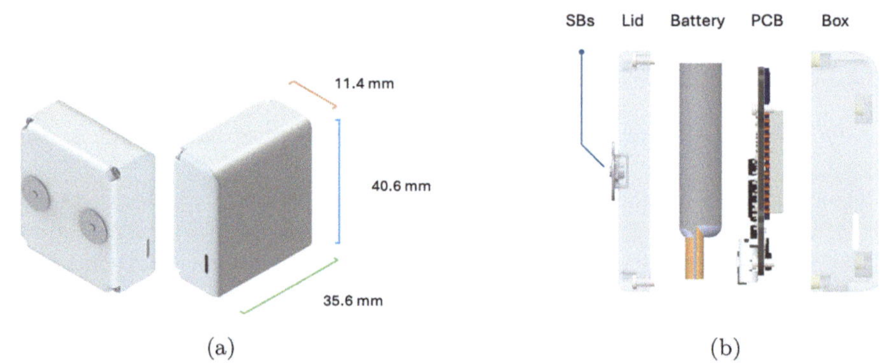

Fig. 6 Device assembly showing **(a)** collapsed view of the device (after assembly) with annotated dimensions and **(b)** exploded view, where the lid and box are rendered as transparent elements to reveal the inserts and screws. SBs refers to the magnetic SBs

Fig. 7 Internal view of the device enclosure, illustrating the screws used to secure the PCB to the inner wall of one of the enclosure parts

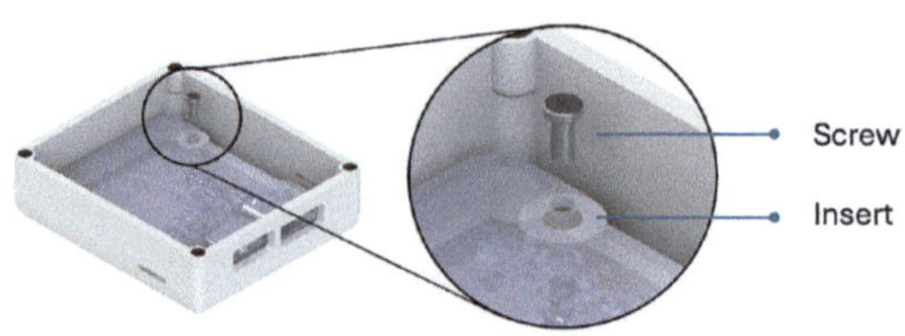

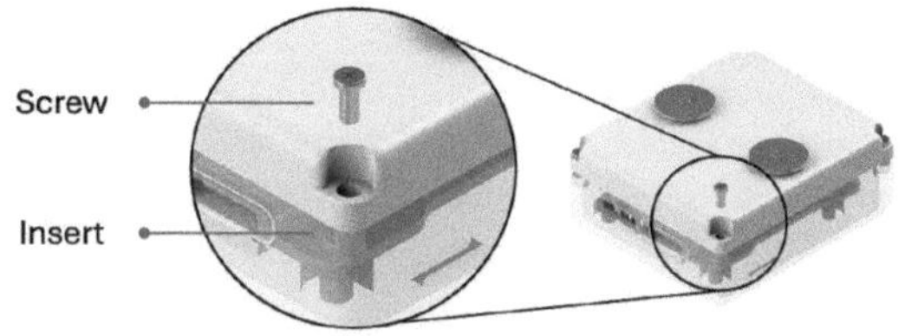

Fig. 8 Enclosure assembly of the device, showing the use of screws to secure the two parts together

4.2 Hardware Optimization

The early use of the ScientISST CORE in experimental settings provided valuable insights, though some limitations were identified, particularly regarding its size and usability—key factors when transitioning towards an end-user product [10]. These limitations included the need for external components and interconnections to the main module, which increased the setup's complexity and time requirements. Although the CORE hardware itself is slim, lightweight, and compact, the integration with sensors and the battery resulted in a bulkier initial prototype. These issues encouraged the optimization of the device's base hardware from its general-purpose configuration (initially provided by the ScientISST CORE) to a more use case-specific design. A significant advantage of the ScientISST CORE is its open-source nature, which has enabled iterative improvements, making it an invaluable tool not only for the initial prototyping stages but also in further stages towards a customized and optimized version and end-user-ready product. This flexibility aligns with the concept of the Technology Readiness Level (TRL), allowing for systematic advancements as the product evolves from functioning in controlled laboratory settings to being viable in real-world applications. Based on the challenges and limitations of our early-stage prototype, we will go through the key improvements that led to the Sympathia Sense board. Table 1 summarizes the main device specifications of Sympathia Sense (after hardware optimization).

Sensor Integration The sensors (and their respective connections to the PCB) that were previously external to the device were integrated into the new hardware design. Integrating the sensors without an increase in size was only possible because of the significant gain in free space achieved through hardware miniaturization. The newly included sensors comprise the adaptive EDA sensor [11] and the tri-axial accelerometer. Besides being shorter, the new connections made between the microcontroller and the sensors are also less prone to interference and noise, as they are now incorporated into the PCB with dedicated ground planes, resulting in improved signal integrity and overall system performance. However, to further minimize noise and interference within the PCB, it was important to implement additional strategies. This included placing digital and analog circuitry as far apart as possible to avoid cross-talk and interference, maintaining appropriate spacing between adjacent and parallel traces, and properly implementing ground and power planes.

Table 1 Device specifications of Sympathia Sense

Biosignals	EDA (1 Ch) + 3-axial ACC (3 Ch)
Sampling rate	1 to 8000 Hz (configurable)
Communication	Bluetooth (range up to 10 m)
Synchronization*	LED or light sensor
Charging port	USB-C
Weight	30 g/1.06 oz
Size	41 × 36 × 11 mm/1.6 × 1.41 × 0.43 in

* Optical synchronization with a second device for alignment of biosignals in post-processing. This is achieved via an auxiliary 12-bit analog input channel or a 1-bit digital I/O line, both available for synchronization purposes

Hardware Miniaturization The result of hardware miniaturization is shown in Fig. 9, and it was achieved through the elimination of unneeded circuitry and the substitution of components by their lowest footprint equivalents. Examples of removed circuitry include RC filters connected to analog input channels which are not used in the new hardware version and pads for the Grove Surface-Mount Technology (SMT) connectors.[1] As for footprint optimization, most resistors and capacitors with a 0603 footprint were substituted with ones of the same value but with a 0201 footprint. Specifically, this footprint substitution results in an area[2] reduction of 86% for each component. Another example of package optimization is the external Analog-to-Digital Converter (ADC) which has a TSSOP-20 footprint in the ScientISST CORE and is substituted by its QFN equivalent package for the new hardware version, resulting in an area reduction of 66%. Overall, selecting smaller footprints and component packages led to more space availability for traces and vias at the outer layers of the PCB, which can be decisive in miniaturizing complex PCB designs.

As shown in Fig. 10b (blue rectangle) the ESP-32 module was retained in the new version. Although further miniaturization of this component was possible, it was not pursued due to the potential time and cost implications. For instance, redesigning the PCB antenna while maintaining the same connectivity range (Wi-Fi and Bluetooth) would involve tuning its pi-network impedance matching circuit, which is a complex procedure. Moreover, while this approach would require recertifying the board for

[1] The Grove interface was developed by Seeed Studio are commonly used in educational and DIY hardware projects https://wiki.seeedstudio.com/Grove_System/.

[2] By area we mean the component's courtyard. In other words, we are referring to the area resulting from the vertical projection of the component on the PCB.

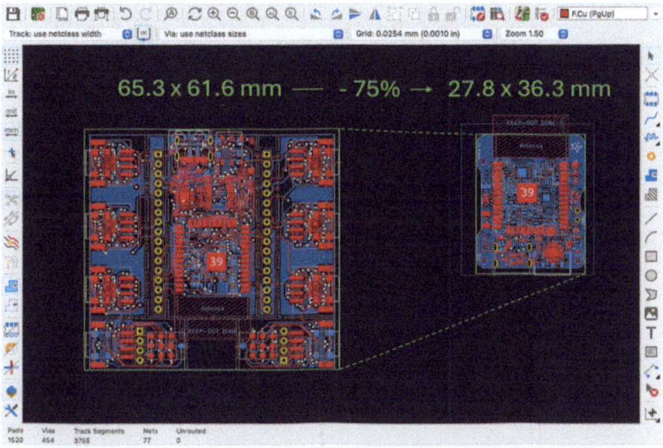

Fig. 9 Overall comparison of the ScientISST CORE and Sympathia Sense PCB dimensions

wireless connectivity, the original ESP-32 module's antenna is already certified for FFC[3] and CE,[4] which is advantageous.

Figure 10a, b show the top and bottom layers of the device's PCB, respectively. In these figures, the colored shapes represent different functional locations within the PCB. The ESP32 microcontroller (blue) is located at the top side within the top layer, close to one of the PCB edges, support circuit and sensors (red), and side (yellow) and front (green) interface components. In these figures, the arrows highlight the direction of the connection for the interface components, which include the battery, detachable connectors, and additional sensors.

4.3 Adaptive EDA Sensing

The adaptive EDA sensor used in the device has been previously validated through comparative EDA acquisitions with a gold standard device [7]. Here we demonstrate how the sensor works. To avoid signal saturation and maintain linearity, the sensor adjusts its configuration by selecting a Digital-to-Analog Converter (DAC) value whenever the dead zone is approached. Following each adjustment, there is an apparent sharp increase (or decrease) in the amplitude of the EDA signal, which can also be reflected in the Electrodermal Response (EDR) signals. These amplitude variations, which we will refer to as transients, are artifacts resulting from the DAC change, and, during these periods, the actual skin conductance is still varying

[3] In the United States, the FCC (Federal Communications Commission) certification ensures that the device complies with regulations for limiting radio frequency emissions and maintaining interference standards.

[4] In Europe, the CE (Conformité Européenne) marking indicates that the device meets the electromagnetic compatibility directive, ensuring it does not cause or receive unacceptable levels of electromagnetic interference.

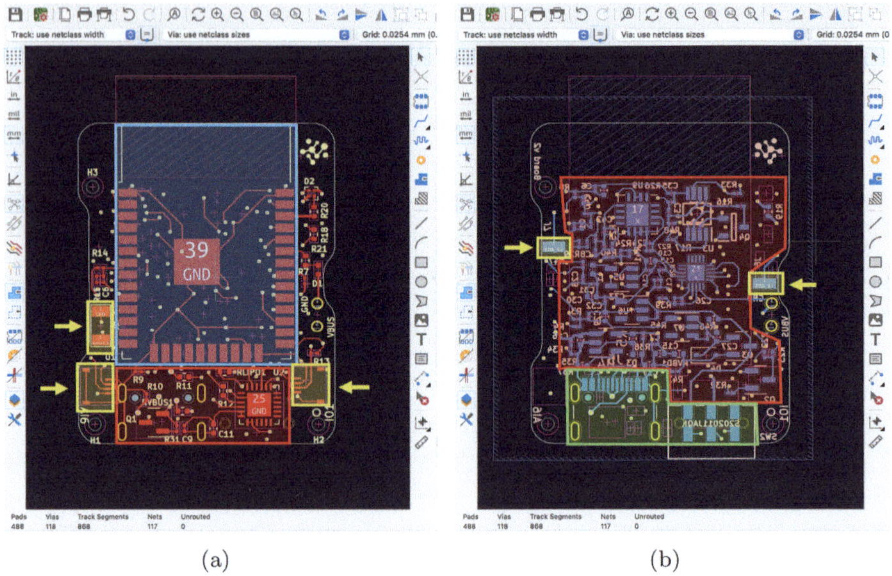

Fig. 10 Device's 4-layer PCB in the PCB design software KiCad (PCB Editor mode). The two inner layers are hidden for visualization purposes. (a) Top layer. (b) Bottom layer

smoothly. These transients are highlighted in gray in the top plot in Fig. 11 and result from the dynamic selection of the transfer functions shown in Fig. 12 via the DAC setting.

To process EDA signals, one can either use the original signals, which include transient artifacts along with the DAC signal, or opt to reconstruct the EDA signals. In the first case, the EDA signals may appear discontinuous or challenging to interpret due to sensor adjustments. However, when using machine learning or deep learning models, this complexity is not a limitation. These models can process both the EDA/EDR signals and the DAC signal together in the feature space, using the DAC information to compensate for amplitude transients, effectively treating them as different representations of the same underlying data. However, if one needs to represent the EDA signal as a continuous time series, then a reconstruction method needs to be employed, as illustrated in the top and bottom plots in Fig. 11.

Figure 13 illustrates the adaptive mechanism of the EDA sensor and the reconstruction of the EDA signal into a continuous time series. The first and second plots at the top present close-up views of the original EDA and the DAC signals, respectively. Notably, between the 10- and 12.5-second marks, the DAC signal decreases from 163 to 162 (time domain). This reduction occurs as the EDA signal nears the lower boundary of the linear zone specified by the sensor configuration at DAC $= 163$. This DAC adjustment induces a transient phase in the EDA signal before its amplitude stabilizes (i.e., after the region labeled "Transition range"). In the transfer function (conductance domain), this adjustment corresponds to the abrupt change shown in the zoomed-in window in Fig. 12. Here, the curve

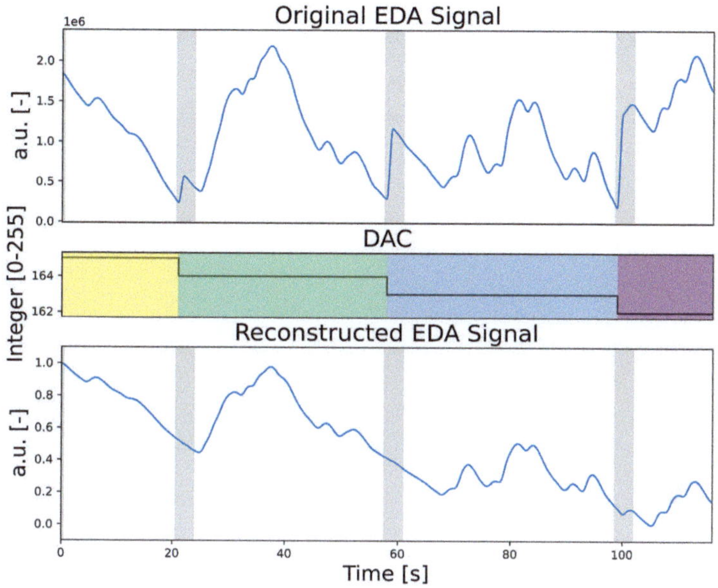

Fig. 11 EDA signal reconstructed into a continuous time series. The original EDA signal is shown at the top along with the DAC signal (first and second plots, respectively). The amplitude transients caused by the change in sensor configuration are highlighted in gray. The reconstructed EDA signal (third plot) shows rather continuous signal segments, which visually preserve the overall decrease in baseline conductance. The DAC signal is overlayed with colored segments to highlight the transitions

representing the sensor configuration DAC $= 163$ (in dark blue) is descended until the lower boundary is reached (1). This event triggers a shift in sensor configuration to DAC $= 162$ (in light blue), represented by the vertical transition (2), which allows the sensor to acquire the EDA signal within the extended linear region (3).

The pseudo-code for the reconstruction algorithm is shown in Algorithm 1. To reconstruct the continuous EDA signal, the algorithm first detects the transition range (1), which corresponds to the transient amplitude variation occurring shortly after the DAC change (red segment in Fig. 13). This is done by identifying the transition point as the maximum derivative of the EDA signal within a window of $[-2, 2]$ seconds centered around the timestamp of the DAC change (red circle labeled transition). Next, the limits of the transition range are determined. The first limit is identified by moving backward from the transition point until the signal's derivative falls below 10% of its value at the transition point, or until the beginning of the transition range is reached. Similarly, the last limit is found by moving forward from the transition point until the derivative is less than 10% of the transition point derivative, or until the end of the transition range is reached. The 10% threshold was determined empirically.

Following this, the algorithm proceeds to the mirroring step (2). Here, the transition range is duplicated, resulting in a preceding range of equal length. The

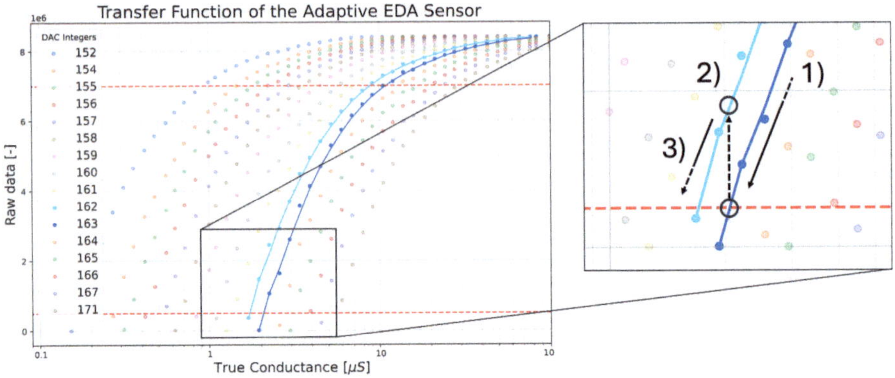

Fig. 12 Transfer functions of the adaptive EDA sensor, showing the curves for DAC = 163 (dark blue) and DAC = 162 (light blue). The zoomed-in view illustrates the adaptive mechanism: as amplitude decreases, DAC = 163 descends to the lower boundary (1), triggering a shift to DAC = 162 (2), allowing continued EDA acquisition within the linear region (3)

signal is then mirrored by copying the segment from the preceding range into the transition range. To ensure continuity, the amplitude of the signal within the preceding range is adjusted by adding the difference between the offset at the end and the beginning of the range to the copied segment. The resulting signal, shown in the third plot of Fig. 13, no longer contains the transient effect, although an amplitude offset remains at its end.

The final step involves amplitude offset adjustment (3). Similarly to the last step in (2), the first and last points of the transition range are identified, and the difference between them (in reverse order) is applied to the signal segment following the transition range. The final reconstructed signal is displayed in the fourth plot of Fig. 13.

5 Proof-of-Concept

To highlight the potential use cases of Sympathia Sense, namely, in affective computing and health monitoring, two exploratory experiments were performed on a healthy subject, while collecting EDA and Accelerometer (ACC) signals through the sock. The first experiment consisted of visualizing an emotional video and the second one was a breath-work exercise. The experiments were performed voluntarily by a 25-year-old healthy male subject, making use of the Sympathia Connect app.

Sympathia Connect Figure 14 shows the Graphical User Interface (GUI) of Sympathia Connect. This app contains several functional blocks highlighted in the figure. Among the available settings, the user selects the device to connect to, and the

Algorithm 1 Algorithm for EDA signal reconstruction

```
Define algorithm parameters
th_val = 0.1 10% for the derivative threshold
window_hs = 2000 Window half size
extra_span_fwd = 200 Extra span for transition region (forward)

Step 1: Detection of the tansition range in EDA signal
Indices where DAC changes (change in sensor configuration)
change_idx = find_changes(DAC)

for idx in change_idx do

    Working window centered around the DAC change
    start_idx = idx - window_hs
    end_idx = idx + window_hs

    EDA_seg = EDA[start_idx : end_idx] Extract segment of the EDA signal
    EDA'_seg = diff(EDA_seg) Compute derivative approx. from the segment
    max_idx = argmax(abs(EDA'_seg)) Obtain index at maximum derivative
    threshold = th_val × abs(EDA'_seg[max_idx]) Compute threshold

    Walk along the signal (from max_idx) until the derivative is lower than threshold
    Backward Search
    start_bwd = idx - 20 Minimum distance from transition
    while start_bwd > 0 and abs(EDA_seg'[start_bwd]) > threshold do
        start_bwd -= 1
    end while

    Forward Search
    % Forward Search
    end_fwd = idx + 20 Minimum distance from transition
    while end_fwd < len(EDA'_seg) and abs(EDA'_seg[end_fwd]) > threshold do
        end_fwd += 1
    end while
    end_fwd += extra_span_fwd Additional span for smoother reconstruction

    Step 2: Mirroring stage
    Compute timstamps of the preceding segment
    prec_start = start_bwd - (end_fwd - start_bwd)
    prec_end = start_bwd

    Copy and mirror the preceding segment into transition range, with amplitude offset correction
    prec_seg = EDA_seg[prec_start : prec_end]
    amp_offset = EDA_seg[prec_end] − EDA_seg[prec_start]
    EDA_seg[start_bwd : end_fwd] = prec_seg + amp_offset

    Step 3: Final amplitude offset adjustment
    final_amp_offset = EDA_seg[end_fwd] − EDA_seg[end_fwd + 1]
    EDA_seg[end_fwd :] = EDA_seg[end_fwd :] + final_amp_offset

    Update EDA_seg in whole EDA_seg signal

end for
```

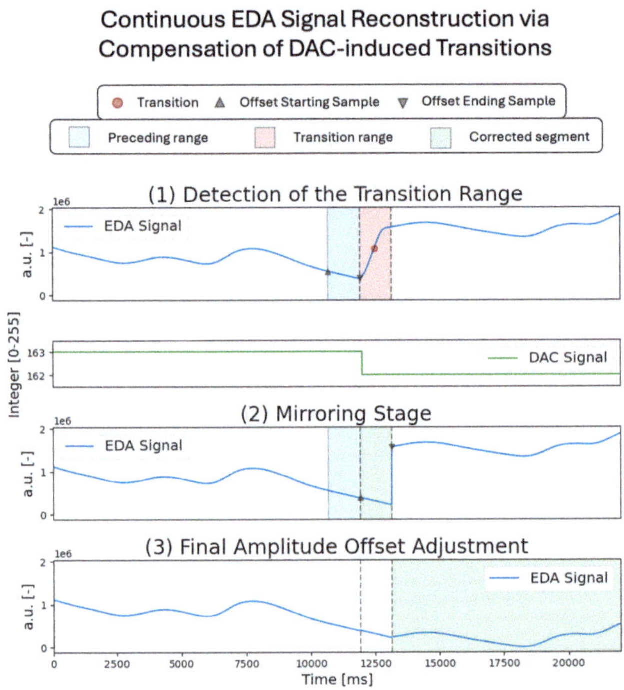

Fig. 13 Filtered EDA and DAC signals with EDA reconstruction through DAC-induced transition compensation. The first two plots show the filtered signals and detection of the transition range (1). The third plot displays the EDA signal after the mirroring step (2), and the fourth plot shows the final reconstructed signal after amplitude offset adjustment (3)

acquisition settings, namely, duration, sampling rate, enabled signal modalities (e.g., EDA and/or accelerometer) and the starting DAC value. Although not used in this experiment, it is also possible to trigger a short LED pulse on the device, allowing to optically synchronize it with a second device. Before starting the acquisition, the user also sets the saving preferences, namely, the saving directory and filename naming method (timestamp-based on incremental), as shown in Fig. 15. When the acquisition starts, the selected signal modalities are plotted in real time, as shown in Fig. 16, which confirms whether the acquisition is running as intended. Finally, the collected signals are saved in real time in a .csv file, which is illustrated in Fig. 17.

Signal Processing The EDA signals were processed by employing standard filtering methods, through the BioSSPy toolbox [12]. The filtering steps are illustrated in Fig. 18, starting from the raw EDA signal. It is worth mentioning that the previously described reconstruction method was pre-applied to the raw signals. This way, the following filtering methods smooth any possible remaining amplitude offsets. EDA signals were filtered using a fourth-order Butterworth low-pass filter with a cutoff frequency $f_c = 5$ Hz and then smoothed using a moving average filter with a window

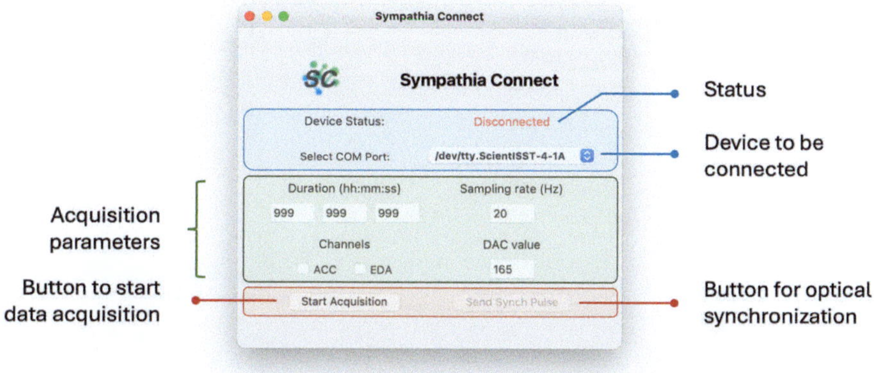

Fig. 14 Sympathia Connect GUI, with the three functional sections. In Section 1 (in blue) the user connects to the device and can check its status. Section 2 (in green) contains the acquisition settings, which are set before the acquisition. These settings include the duration of the acquisition (hours, minutes, and seconds), the sampling rate (Hertz), the signal modalities which are enabled (EDA and/or accelerometer), and the starting DAC value. Finally, Section 3 (in red) has a button to start the acquisition and another to trigger a synchronization pulse (optional). By triggering this button, the device emits a 1-second LED pulse, which can be detected by a second device for optical synchronization

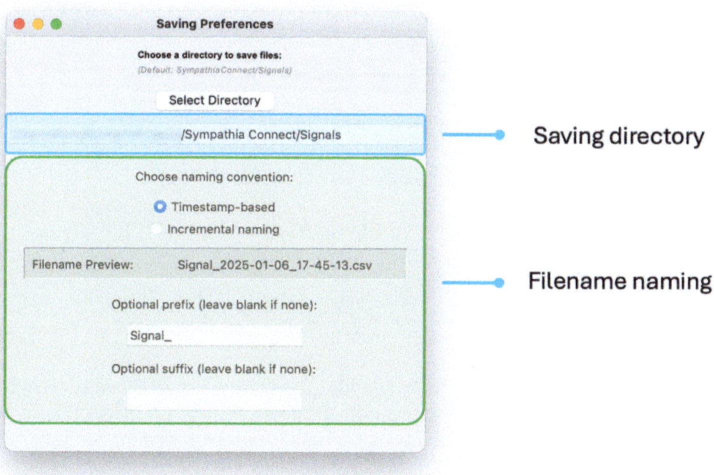

Fig. 15 Saving preferences window. The user can select the directory path where the acquisition data is saved and customize the naming method for the saved files. The files are saved in the .csv format

size of 750 ms. EDR signals were obtained by adding an additional filtering stage in between, a fourth-order Butterworth high-pass filter with a cutoff frequency $f_c = 0.05$ Hz.

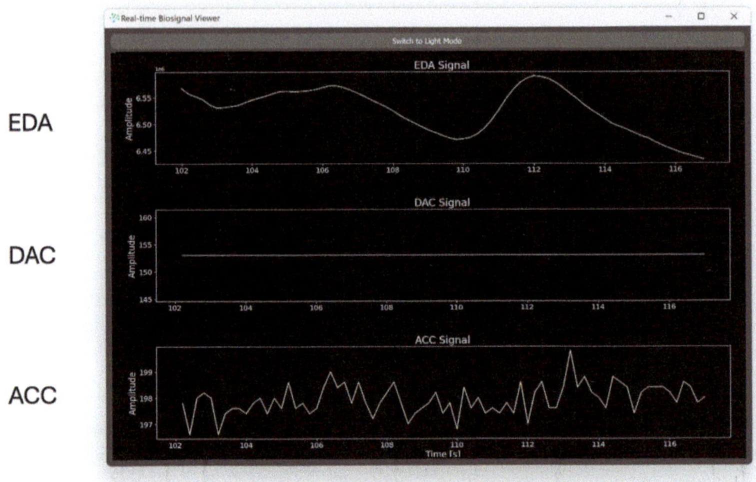

Fig. 16 Real-time biosignal visualizer. In this example, both the EDA and the ACC modalities are enabled. The DAC signal is always displayed (and stored) when the EDA is enabled. Moreover, the DAC is constant because the EDA signal being measured is within the linear region (constant sensor configuration). For simplification, the 3-axial ACC signals are computed and represented as a single time series, the vector magnitude ($VM = \sqrt{acc_X{}^2 + acc_Y{}^2 + acc_Z{}^2}$)

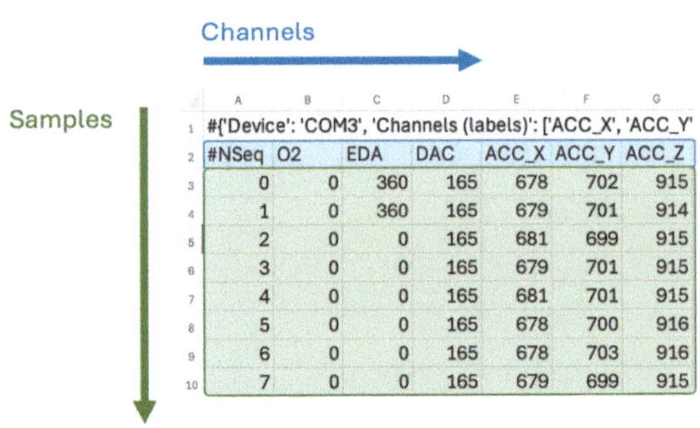

	A	B	C	D	E	F	G
1	#{'Device': 'COM3', 'Channels (labels)': ['ACC_X', 'ACC_Y'						
2	#NSeq	O2	EDA	DAC	ACC_X	ACC_Y	ACC_Z
3	0	0	360	165	678	702	915
4	1	0	360	165	679	701	914
5	2	0	0	165	681	699	915
6	3	0	0	165	679	701	915
7	4	0	0	165	681	701	915
8	5	0	0	165	678	700	916
9	6	0	0	165	678	703	916
10	7	0	0	165	679	699	915

Fig. 17 Structure of the saved acquisition data file. The second line of the header (in blue) contains the labels of the recorded signals. The column labelled as NSeq corresponds to the sequence transmitted by the device and is used to check for data loss. The remaining columns correspond to the acquired signals: LED state (optical synchronization), the EDA and DAC signals, and the 3-axial acceleration signals, ACC_X, ACC_Y, ACC_Z. The lines in green correspond to the collected samples at each timestamp. The difference in time between consecutive samples is the sampling period ($\Delta T = 1/f_s$ [s])

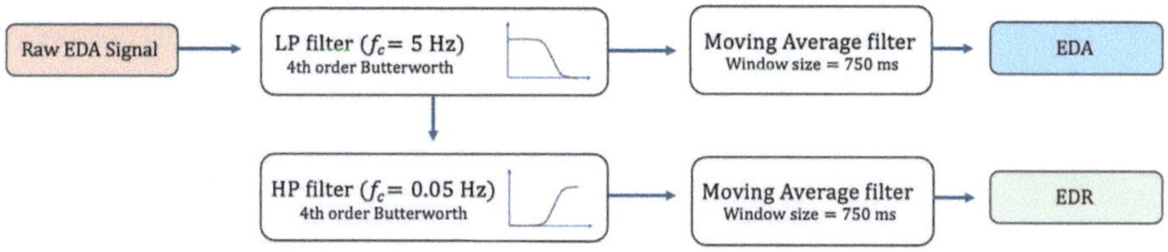

Fig. 18 Illustration of the signal processing steps to obtain the EDA and EDR signals from the raw EDA signals

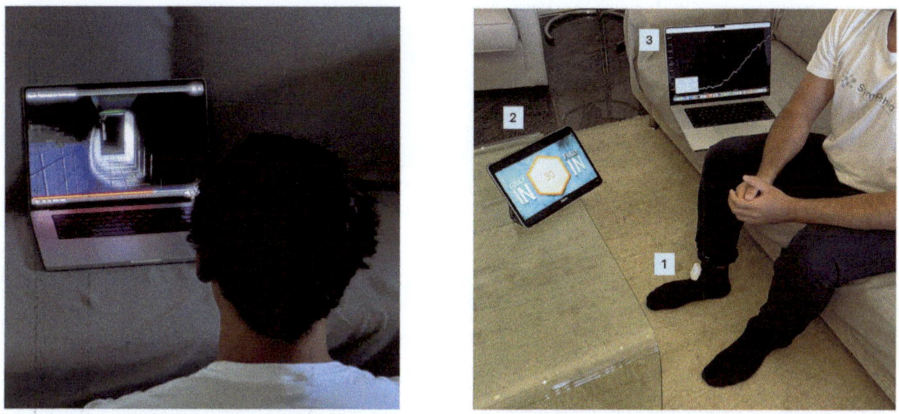

(a) (b)

Fig. 19 Experimental setup for the proof-of-concept acquisitions. **(a)** Visualization of the emotional video under the dark to amplifies fear responses during the session. **(b)** WHBM session: (1) Sympathia Sense, (2) tablet displaying the breathing instructions, and (3) laptop receiving the EDA and ACC signals in real time, through the Acquisition App

Emotional Video As illustrated in Fig. 19a, in the first experiment, the subject watched the game-play of Slender,[5] which is a first-person survival horror video game. The resulting EDA and ACC signals are shown in Fig. 20, along with the EDR, extracted in post-processing from the EDA signal, and the DAC signal, which is used for controlling the EDA sensor. The video is 14 minutes and 47 seconds long (887 seconds) and is synchronized with the signal acquisition performed through the device. The participant promptly annotated whenever he experienced fear by pressing a key on the keyboard, resulting in the 4 "stimuli" marks in the EDA and EDR plots. As shown, near timestamps from 0 to 2, the EDR signal displays significant phasic variations, which is associated with sweat release induced by the response to fear. Following the same reasoning, there is a significant increase in the baseline of the EDA signal after these timestamps, highlighting the increase in

[5] From legend5926's YouTube channel: https://www.youtube.com/watch?v=E9wBelF9d-o (accessed on August 21, 2024)

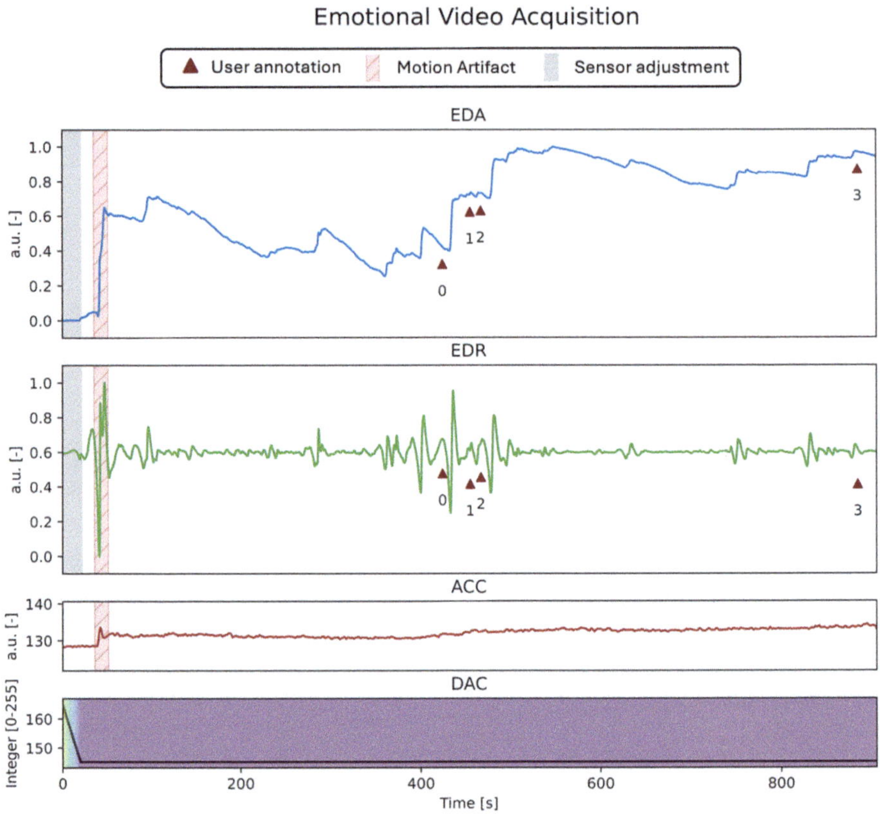

Fig. 20 Signal acquisition during the visualization of an emotional video. The following plots are shown (from top to bottom): EDA, EDR, ACC, and DAC signals. The plots are annotated. Stimulus (red triangles) corresponds to the manual annotations performed by the user during the video session. Motion artifacts (red ranges) correspond to artifacts on the EDA and EDR signals induced by variations in body movement, as confirmed by the ACC signals. Sensor adjustment (gray ranges) are periods of change in the sensor configuration, as confirmed by the variation in the DAC signal

the tonic component of the signal. It is worth noting that these observations are not affected by artifacts resulting from changes in sensor configuration or movement; these artifacts only occur in the first portion of the acquisition, as shown in the red and gray sections, respectively.

Breath-Work Exercise In the second experiment, the subject voluntarily watched a video of the WHBM[6] and performed the breath work, following the instructions

[6] From Wim Hof's YouTube channel: https://www.youtube.com/watch?v=tybOi4hjZFQ (accessed on August 21, 2024)

Table 2 Instructions of the video "Guided Wim Hof Method Breathing," performed by the subject

Phase	Time interval	Duration	Instruction description
P0	0 to 60s	60s	Normal breathing (physiological baseline)
P1	60 to 150s	90s	Series of 30 deep breaths at a fast cadence
P2	150 to 210s	60s	Deep breath-in, followed by a 1-minute hold period
P3	210 to 225s	15s	Release, followed by a deep breath and a 15-second hold period
P4	225 to 300s	75s	Breathing normally (physiological baseline)

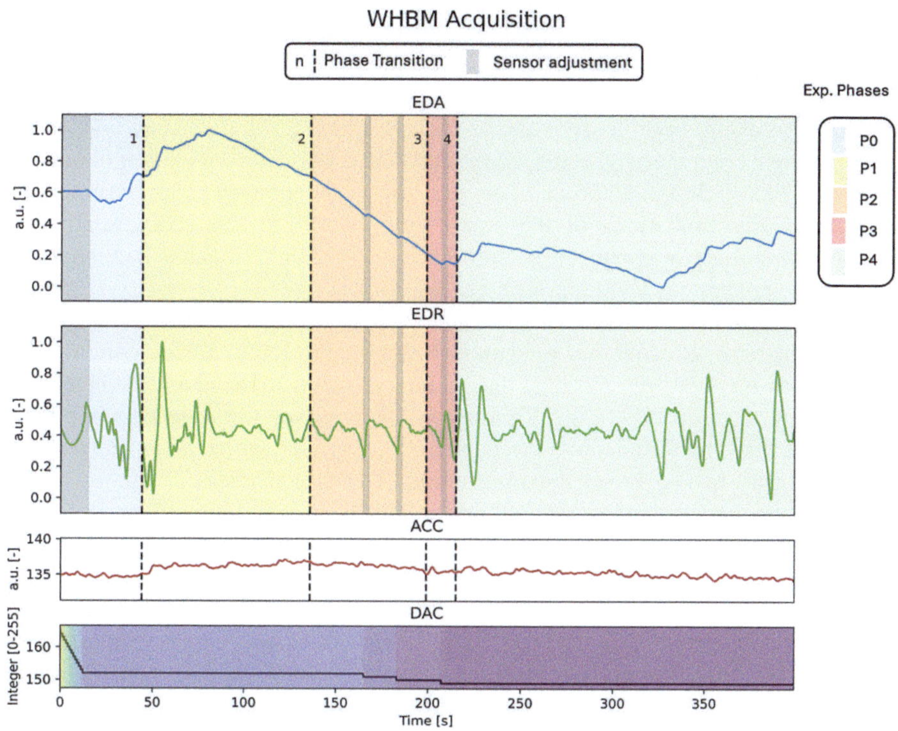

Fig. 21 WHBM signal acquisition. The plots follow the same order as in Fig. 20. Each phase of the breath work, from P0 to P4, is highlighted in a different color and is separated by a vertical dashed line denoting its start in time

in the video, which are listed in Table 2. This experimental setup is illustrated in Fig. 19b. It is worth mentioning that breath-work techniques, especially fast-paced practices such as this one, are known to induce peripheral sweating through the activation of the SNS [13, 14].

Similarly to the previous experiment, the obtained signals are shown in Fig. 21. In this figure, each phase, from P0 to P4 (from Table 2), is shown in a different color, and each transition is represented by a dashed line. The results support that it was possible to observe an increase, especially in the tonic component of the EDA signal during phase P1 (yellow range), which is in line with previous research upon SNS-inducing breath work [15].

6 Summary

Sympathia Sense represents a significant advancement in wearable technology for physiological monitoring, combining multiple domains of engineering, from e-textiles to signal processing. Moreover, it provides yet another example of a potential application of smart socks. This book chapter illustrates the successful transition from an initial, general-purpose prototype to a highly specialized device for physiological monitoring, namely, electrodermal activity (EDA) and body movement sensing. For this, an e-textile sock was first developed using conductive yarns, through computerized embroidery. Then the acquisition device was optimized through iterative design improvements and sensor integration. This optimization addressed early limitations of the smart sock related to size, complexity, and reliability, resulting in enhanced performance and usability, making it suitable for real-world applications and long-term acquisitions.

This chapter also revisits the adaptive EDA sensor, previously validated in [7, 11], and demonstrates its capability to acquire EDA signals across a wide conductance range (0.1 to 100 μS) with minimal discontinuity, thanks to the newly implemented signal reconstruction algorithm. This algorithm removes the transient amplitude variations caused by the adaptive mechanism of the sensor, which helps maintaining signal integrity over a broad dynamic range. The effectiveness of this approach ensures the reliability of the collected data, whether analyzed directly or processed using advanced machine learning models.

The proof-of-concept experiments showcase the potential of Sympathia Sense in practical applications. The device successfully captured EDA and accelerometer (ACC) signals during an emotional video and a breath-work exercise, and the signals align well of amplitude variation in the EDA signals. The results support that Sympathia Sense is able to measure the activation of the SNS through the presentation of emotional stimuli (video) and self-induced physiological changes (breath work). Overall, Sympathia Sense exemplifies the integration of existing open-source and adaptable hardware and software tools to accelerate research in various fields, such as affective computing and health monitoring, moving a step further towards into a real-world wearable product. In conclusion, the following key points highlight the main contributions and findings presented in this chapter:

- It is possible to **develop and manufacture customizable e-textiles** for biosignal sensing, namely, through computerized embroidery.
- Early hardware design limitations, such as size, complexity, and reliability, should be addressed through **iterative optimization and sensor integration**.
- It is possible to use **off-the-shelf components** to produce the **hardware enclosure**, namely, through 3D modelling and printing.

- **Adaptive sensing mechanisms** can be used in sensors to acquire biosignals over a broader measurement range, but may require implementing customized signal processing techniques for further processing and visualization purposes.
- Sympathia Sense demonstrates the implementation of a smart sock for electrodermal activity (EDA) and body movement monitoring, making use of **open-source and adaptable hardware and software**.
- Sympathia Sense aims at **accelerating research** in fields such as affective computing and health monitoring and moving towards a market-ready wearable product.

Acknowledgments This work was partially funded by FCT/MECI through national funds and when applicable co-funded EU funds under project UID/50008: Instituto de Telecomunicações. It was also co-supported by the European Regional Development Fund (FEDER), through the Lisbon Regional Programme (LISBOA 2030) of the Portugal 2030 framework—LISBOA2030-FEDER-01318700 (ComSense)—and by the FCT PhD grant 2024.02772.BDANA. The authors would like to thank the Pattern and Image Analysis group of IT-IST, including all its students and researchers, for their valuable support throughout this work.

References

1. W. Boucsein, *Electrodermal Activity* (Springer, Berlin, 2012)
2. A. Greco, G. Valenza, E.P. Scilingo, *Advances in Electrodermal Activity Processing with Applications for Mental Health* (Springer, Berlin, 2016)
3. A.F. Payne, A.M. Schell, M.E. Dawson, Lapses in skin conductance responding across anatomical sites: comparison of fingers, feet, forehead, and wrist. Psychophysiology **53**(7), 1084–1092 (2016)
4. P.J. Bota, C. Wang, A.L. Fred, H.P. Da Silva, A review, current challenges, and future possibilities on emotion recognition using machine learning and physiological signals. IEEE access **7**, 140990–141020 (2019)
5. C. Matti, S. Essig, Z. Föhn, A. Balthasar, The role of wearable sensors in the future primary healthcare–preferences of the adult Swiss population: a mixed methods approach. J. Med. Syst. **47**(1), 111 (2023)
6. G. Wu, S. Siegler, P. Allard, C. Kirtley, A. Leardini, D. Rosenbaum, M. Whittle, D.D. D'Lima, L. Cristofolini, H. Witte, et al., ISB recommendation on definitions of joint coordinate system of various joints for the reporting of human joint motion—part I: ankle, hip, and spine. J. Biomech. **35**(4), 543–548 (2002)
7. A.F. Ferreira, H.P. da Silva, H. Alves, A. Fred, Experimental evaluation of an adaptive gain sensor for electrodermal activity monitoring in a smart sock form factor, in *Proceedings of the 2024 IEEE 22nd Mediterranean Electrotechnical Conference (MELECON)* (IEEE, 2024), pp. 779–784
8. J. Stanley, J.A. Hunt, P. Kunovski, Y. Wei, A review of connectors and joining technologies for electronic textiles. Eng. Rep. **4**(6), e12491 (2022)
9. A. Fortes Ferreira, H. Alves, H.P. da Silva, N. Marques, A. Fred, Exploring the electrical robustness of conductive textile fasteners for wearable devices in different human motion conditions. Sci. Rep. **14**(1), 7872 (2024)
10. A.F. Ferreira, H.P. da Silva, H. Alves, N. Marques, A. Fred, Feasibility of electrodermal activity and photoplethysmography data acquisition at the foot using a sock form factor. Sensors **23**(2), 620 (2023)

11. A.R. Banganho, M.B. dos Santos, H.P. da Silva, Design and evaluation of an electrodermal activity sensor (EDA) with adaptive gain. IEEE Sensors J. **21**(6), 8639–8649 (2021)
12. P. Bota, R. Silva, C. Carreiras, A. Fred, H.P. da Silva, BioSPPy: a python toolbox for physiological signal processing. SoftwareX **26**, 101712 (2024)
13. O. Muzik, K.T. Reilly, V.A. Diwadkar, "brain over body"—a study on the willful regulation of autonomic function during cold exposure. NeuroImage **172**, 632–641 (2018)
14. M. Kox, L.T. Van Eijk, J. Zwaag, J. Van Den Wildenberg, F.C. Sweep, J.G. Van Der Hoeven, P. Pickkers, Voluntary activation of the sympathetic nervous system and attenuation of the innate immune response in humans. Proc. Natl. Acad. Sci., **111**(20), 7379–7384 (2014)
15. A.F. Ferreira, Novel wearable sock for multimodal peripheral biosignal acquisition: towards unobtrusive seizure monitoring in epilepsy. Master's Thesis, Instituto Superior Técnico, 2022

Case Study: COPD Pulmonary Rehabilitation with Biofeedback Physiotherapy

Inês Carvalho, Ana Sofia Cacais do Carmo, Mariana Abreu, Patrícia Justo Bota, Fátima Rodrigues, Ana Martins, Rute Roberto, Joana F. Pinto, Gonçalo Chambel, and Hugo Plácido da Silva

1 Introduction

Chronic Respiratory Diseases (CRDs) represent a significant global health challenge, contributing substantially to both mortality and disability worldwide. According to the World Health Organization (WHO),[1] beyond imposing a substantial

[1] WHO topic on CRDs available at https://www.who.int/health-topics/chronic-respiratory-diseases (last access on 11-2023)

I. Carvalho (✉) · A. S. C. do Carmo · M. Abreu · P. J. Bota · H. P. da Silva
Instituto Superior Técnico, Universidade de Lisboa, Lisbon, Portugal

IT - Instituto de Telecomunicações, Lisbon, Portugal

F. Rodrigues
Pulmonary Rehabilitation Unit. Thoracic Department, North Lisbon University Hospital Centre, Lisbon, Portugal

Institute of Environmental Health, Associate Laboratory TERRA, Lisbon Medical School, Lisbon University, Lisbon, Portugal

A. Martins
Pulmonary Rehabilitation Unit. Thoracic Department, North Lisbon University Hospital Centre, Lisbon, Portugal

Higher School of Health of the Portuguese Red Cross, Lisbon, Portugal

R. Roberto
Pulmonary Rehabilitation Unit. Thoracic Department, North Lisbon University Hospital Centre, Lisbon, Portugal

J. F. Pinto · G. Chambel
Clynx Health, Clynxio LDA, Lisbon, Portugal

H. P. da Silva et al. (eds.), *Open Source Biomedical Engineering*,
https://doi.org/10.1007/978-3-032-03655-1_12

societal burden, CRDs profoundly impact the lives of those affected, leading to symptoms such as shortness of breath, chest pain, and fatigue.

Among CRDs, COPD stands out as one of the most prevalent and its numbers continue to rise. The WHO predicts that COPD will become the third most common cause of death by 2030, affecting approximately 64 million people globally [1]. Moreover, it often remains underdiagnosed and undertreated in many countries [2, 3].

Fortunately, treatments such as Pulmonary Rehabilitation (PR) have demonstrated their effectiveness in reducing symptoms, slowing disease progression, and enhancing daily functioning [4, 5]. The main principles of PR aim to minimize symptom burden and enhance exercise capacity, maximizing the patient's independence [6].

Telerehabilitation, leveraging digital technology, has the potential to enhance the uptake and accessibility of rehabilitation services. Studies in COPD patients have demonstrated that telerehabilitation can yield outcomes similar to those of traditional center-based PR, with no identified safety issues [7].

One of the most desired features in a telerehabilitation system is the measurement of physiological processes such as respiration. Telerehabilitation acknowledges that patients need do not follow a "one-size-fits-all" approach, offering a patient-centered solution [3], which is particularly important in the context of COPD treatment. Setting up a telerehabilitation solution should be adapted to the applied context.

This study aimed to explore the feasibility, usability, perceived impact, and user experience with a set of exemplar exercises using a gamified telerehabilitation system[2] enhanced with cardiorespiratory biofeedback, which included the use of wearable technology, for PR. Additionally, the focus was on the specific biofeedback components used, allowing future iterations and refinements to incorporate user-centered viewpoints.

The remainder of the work is organized as follows: Sect. 2 describes the inclusion of gamification, wearables, and sensors, enabling biofeedback in telerehabilitation; Sect. 3 explains the rationale for including cardiorespiratory biofeedback in the context of adapting an existing telerehabilitation system to a Pulmonary Telerehabilitation (PTele) context. Section 4 details the conducted study, including the materials used and the methodology employed, also providing clarification on the implemented biofeedback. Section 5 summarizes the main results from this pilot study. Section 6 provides an in-depth interpretation of the study's results, clarifying the main findings and offering suggestions for future research and recommendations for future biofeedback research in the context of telerehabilitation. Finally, Sect. 7 provides brief conclusions and main findings.

[2] A telerehabilitation system by Clynx: https://clynx.io/ (last access on September 2025)

2 Related Work

The integration of gamification and portable monitoring solutions is a technological trend used in telerehabilitation.

Gamified physiotherapy, also known as serious games, uses technology to provide rehabilitation in engaging and gamified scenarios that can facilitate a "shift in attention," effectively distracting patients from negative sensations such as fatigue and dyspnea during physical activity. This offers patients the opportunity to experience an interesting world and greatly enhance their motivation to exercise [1, 8], which is a cornerstone of PR exercise training. Tey et al. [9], Shih et al. [10], and Colombo et al. [11] provide examples of the use of gamification in the context of breathing retraining and PR.

Wearables and sensors have gained widespread acceptance as promising tools for training and the recovery of motor and coordination functions, enabling professionals to provide real-time support [12, 13]. In addition to motor monitoring, these devices can also facilitate assessing and monitoring various patient parameters, including essential physiological functions such as cardiac and respiratory function [2, 14]. The integration of these tools has opened the doors to unsupervised and adaptive therapy, along with the automation of individualized feedback. Notable features include real-time alerts, exercise adaptation, and motivational cues.

Biofeedback is a well-established technique in rehabilitation designed to improve treatment outcomes by educating and engaging patients. Its integration into telerehabilitation has shown promising potential to enhance engagement and improve outcomes [15]. In designing such strategies, researchers discuss the use of different feedback components (i.e., the mode, content, frequency, and timing of the feedback) in their systems and highlight the need to provide a clear rationale for selecting them. To evaluate the efficacy of the feedback and justify its inclusion, exclusion, or modification in future iterations, researchers commonly assess it using close-ended questionnaires [14].

However, it is important to note that the integration of these technologies into telerehabilitation systems, particularly for monitoring vital signs and other physical biometric data, is still limited. Currently, most telerehabilitation systems focus on physical rehabilitation and do not routinely collect or utilize other vital signs in their delivery [16]. Nevertheless, there are emerging systems and studies in the telerehabilitation field incorporating biofeedback on vital signs. Research on respiratory biofeedback is up and coming, as it may help reconsider the role of monitoring devices in telerehabilitation solutions. For instance, [9, 10, 17–19] have introduced digital biofeedback devices and monitoring techniques with relevance for PTele.

3 Implementation Rationale

Given the large number of patients who undergo PR, such as COPD patients, there is an interest in leveraging technological advancements and telerehabilitation systems to make exercise training more enjoyable, interactive, and automated. The current Clynx Platform (CLYNXIO LDA)[2] is a gamified telerehabilitation solution designed for musculoskeletal physiotherapy.

To adapt it to be more suitable for these patients, the Clynx team met with healthcare professionals, physicians, and Physiotherapists (PTs) specialized in PR, forming a multidisciplinary team to establish requirements and enhancements to address the needs of this physiotherapy rehabilitation context. Opinions and statements of four PTs were formally gathered by applying a set of questions, using a 1–5 Likert scale.

There was a consensus that incorporating cardiac and respiratory monitoring would be an important feature for effectively managing exercise performance in patients undergoing PR, corresponding to a score of 4.5/5. Both Breathing Rate (BR) and Heart Rate (HR) were considered adequate to be used in PR physiotherapy, corresponding to scores of 4.75/5 and 4.25/5. Moreover, besides BR and HR monitoring, two PTs indicated that "Peripheral O2 saturation and symptom assessment (such as shortness of breath and fatigue)" were also considered relevant vital metrics. Another notable enhancement discussed was the integration of guidance for correct Breathing Exercise Coordination (BEC) within the system, which was considered pertinent (scored 4/5). For example, there could be guidance on pacing arm lifts during the inspiratory or expiratory phase. In clinical practice, BEC is integrated as breathing exercises and/or breathing retraining. These involve breathing control techniques and are prescribed to enhance the strength and endurance of respiratory muscles [20].

All PTs fully agreed (scored 5/5) that coordinating breathing and exercise motion is a challenge for the patients. Ultimately, all PTs acknowledged that clinical practice would benefit from automating patient monitoring and providing education on BEC.

4 Materials and Methods

The Ethics Committee of the Centro Academico de Medicina de Lisboa (CAML) approved this study (referenced under the number 140/23). The study adhered to the principles outlined in the Declaration of Helsinki [21], ensuring participants' well-being. Each participant was provided with the choice to participate and, upon their agreement, underwent a thorough explanation of the study's procedures before signing the consent form. This study entailed a proof-of-concept using an existing gamified telerehabilitation solution enhanced with cardiorespiratory biofeedback for PR exercise training of COPD patients, supervised by PTs. Data was collected by a depth camera, wearable sensors, and questionnaires were used to assess users' experience.

Table 1 Patient's anthropological, demographic, pathological details, and oxygen supplementation status. Abbreviations: P = participant. sex = male (M) or female (F), age = age in years, Th. P. = thoracic perimeter in centimeters (cm), pathology description = clinical classification of COPD or other conditions, oxygen sup. = oxygen supplementation in liters per minute (L/min) Pathology description followed the GOLD 2023 guidelines [22]

P.	Sex (M/F)	Age (Years)	Th. P. (cm)	Pathology Description	Oxygen sup. (L/min)
P1	F	52	107	COPD GOLD IV E + ACOS	1 L/min
P2	M	71	88	COPD GOLD I B + Alpha 1 antitrypsin deficiency	–
P3	M	60	101	COPD II B + Alpha 1 antitrypsin deficiency	–
P4	M	67	88	COPD GOLD IV E	4 L/min
P5	M	75	95	COPD GOLD II E	4 L/min
P6	M	73	110	COPD GOLD IV E + Hypercapnic respiratory failure	4 L/min
P7	M	70	95	COPD GOLD III E	–
P8	M	65	97	COPD GOLD III	–
P9	M	64	94	COPD GOLD II A	–
P10	M	78	95	COPD GOLD III A	–

4.1 Participants

A total of 10 patients diagnosed with COPD were recruited from the PR Unit of Hospital Pulido Valente (HPV). The sample consisted of 1 female and 9 males with an average age of (66.3±6.7) years, whose details can be found in Table 1.

All of the 10 patients reported not having used gamification physiotherapy before, while 6 out of 10 had never used wearable Electrocardiography (ECG) and respiration monitoring devices during a rehabilitation physiotherapy session. All respondents reported having some difficulties in maintaining motivation throughout exercise training. Nine of ten mentioned having difficulties in focusing on BEC.[3]

4.2 Cardiorespiratory Biofeedback

The implemented cardiorespiratory biofeedback focused on cardiac and respiratory monitoring, presented with a custom Graphical User Interface (GUI), and guiding BEC, using a visual cue, as further described below.

[3] Supported by preliminary questions posed to the patients.

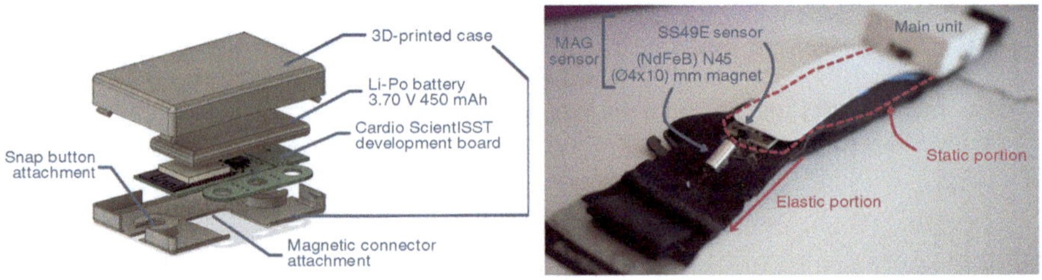

(a) (b)

Fig. 1 Overall hardware constituents. (Adapted from [24]). (**a**) Cardio ScientISST. (**b**) Polar chest strap, magnetic breathing sensor, and ScientISST main unit

4.2.1 Cardiorespiratory Monitoring

Real-time cardiorespiratory monitoring was achieved through the use of a wearable chest strap with embedded Electrocardiography (ECG) and Respiration (RESP) sensors and acquisition unit. A Cardio ScientISST development board [23] (Chap. 2) was used as the main acquisition unit (as illustrated in Fig. 1a), which also included an embedded ECG sensor, interfaced with a Polar Soft Strap (Polar Electro Oy 2023). Respiration monitoring was added through a magnetic field-based sensor, proposed and validated in [24] (shown in Fig. 1b).

Heart Rate (HR) and Breathing Rate (BR) were computed the ECG and RESP signals, respectively. The computation of the HR involved the use of the Hamilton Segmenter [25] for R-peak detection in the ECG QRS complex, utilizing the `BioSSPy` [26] (Chap. 4) Python package. For the BR, the Automatic Multiscale-Based Peak Detection (AMPD) algorithm [27] was employed for peak detection in an adaptive window.

These values were then iteratively displayed on a second screen with a custom GUI developed in Python 3.10.

4.2.2 Breathing Phase Biofeedback

A slightly different version of the game application, referred to as the "breathing-version," was developed as an extension of the commercially available Clynx telerehabilitation solution. In addition to the original game version, it incorporates Breathing Phase Biofeedback (BPB) to guide users on how to coordinate their breathing phases (i.e., inhale and exhale) during the exercise. This integration was customized for each exercise and was developed in collaboration with PTs from

Fig. 2 Visual cue used to give breathing phase feedback

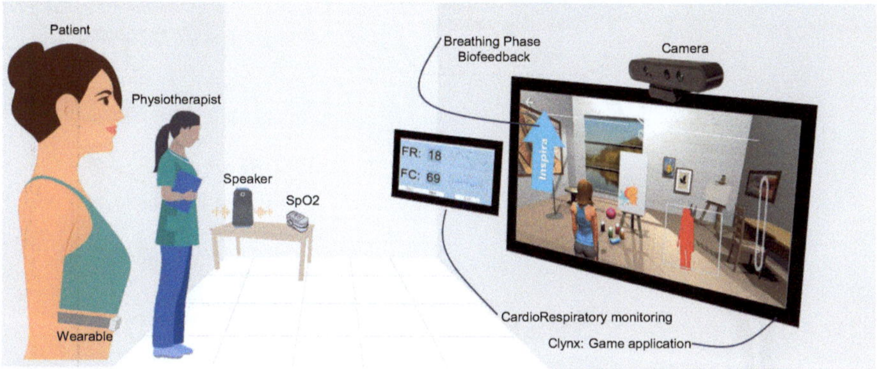

Fig. 3 Experimental setup using two computers, one dedicated to cardiorespiratory monitoring and the other to the Clynx Platform game application

the PR Unit at HPV, drawing upon their valuable insights and expertise. This implementation primarily relied on motion and joint angles, which are provided by a depth camera. When maximum joint angles were detected, it triggered a change in the recommended breathing phase, and users were provided with a graphical cue, as shown in Fig. 2, to assist them with their BEC.

4.3 Experimental Setup

The experimental setup, as shown in Fig. 3, consisted of participants being equipped with the wearable. This wearable was used in conjunction with the commercially available Clynx Platform,[4] which incorporated a depth camera and the game application.

[4] Clynx Platform is available at https://www.clynx.io/pt.html (last accessed on September 2025)

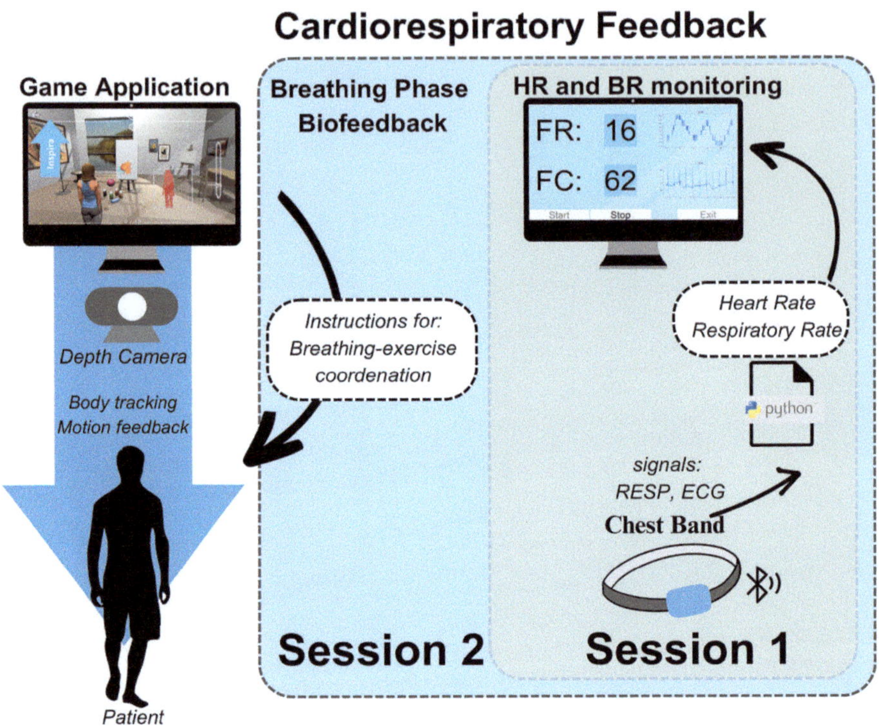

Fig. 4 Cardiorespiratory biofeedback applied in session 1 and in session 2

Two computers were dedicated to the experiment: one served as the display for the game application, while the other facilitated real-time cardiorespiratory monitoring, specifically tracking HR and BR.

To enhance the immersive interaction with the game application, a speaker was incorporated into the setup. Additionally, patients made use of a pulse oximeter, a recommended tool in PR exercise training of COPD patients [28].

4.4 Experimental Procedure

The study design aimed to assess the feasibility and utility of two primary objectives: delivering BPB and enabling real-time cardiorespiratory monitoring using a wearable device. Throughout the study, participants were supervised by PTs.

As detailed in Fig. 4, patient participation in the study was divided into two parts:

- Session 1: patients used the "original" Clynx game application without BPB.
- Session 2: patients used the Clynx game application with instructive BPB on breathing coordination (described in Sect. 4.2.2 as "breathing version").

Table 2 Exercise protocol with the corresponding difficulty and breathing cue descriptions

Code	Exercise name	Difficulty	Repetitions	Breathing cues
EF	Elbow flexion	0°–140° (level 3)	12	Inhale on descending Exhale on ascending
SF	Shoulder flexion	70°–100° (level 3)	12	Inhale on ascending Exhale on descending
SA	Shoulder abduction	40°–50° (level 3)	12	Inhale on ascending Exhale on descending
LA	Leg adduction	40°–50° (level 3)	10	Inhale on descending Exhale on ascending

This approach with separate sessions was intended to create a study design specific enough to distinguish the effect of the feedback from the overall system's impact. In both sessions, additional information regarding HR and BR was displayed on a separate computer screen for the PTs' reference.

In both sessions, patients were instructed to complete a series of exercises, each comprising various motions, including Elbow Flexion (EF), Shoulder Flexion (SF), Shoulder Abduction (SA), and Leg Abduction (LA), using their left limbs in the first session and their right limbs in the second session. At the beginning of each exercise, the PTs explained how to coordinate breathing with the exercise, providing no further guidance during the exercise. Detailed exercise descriptions can be found in Table 2.

During each session, a Case Report File (CRF) was completed by the clinician to provide a brief description of the patient's condition, anonymizing the patient with an assigned Identification Code (ID). Additionally, it was used to record the patient's self-reported perceived physical effort and respiratory discomfort at the end of each exercise, using the modified Borg scale [29]. The clinician assessed the patient's ability to coordinate exercise and breathing for each completed exercise by classifying, with the Likert scale (1–5) [30], the following statement:

Patient was able to coordinate exercise and breathing

At the end of all activities, patients were asked to complete both quantitative and qualitative questionnaires. This approach, as also described by other authors [31, 32], was intended to provide valuable insights and allow for a user-centered design approach to evaluating this biofeedback system.

The questionnaires used in the study comprised several components. Firstly, the System Usability Scale (SUS) was employed to evaluate the overall usability of the enhanced system, which included wearable use, real-time monitoring, and BPB [33]. Additionally, a specific set of questions was applied to assess different aspects of the system, such as the visual feedback component used to provide the BPB in the "breathing-version" game application. These questions generated scores on a 5-point Likert scale and allowed to assess the perceived impact of the system and the subjective quality of enhancement. Furthermore, participants were encouraged to offer any additional comments or statements regarding their experiences.

Finally, PTs were asked to provide their insights about the enhanced system using a similar questionnaire to the one applied to the patients, with additional questions regarding BR and HR monitoring.

4.5 Data Analysis

The SUS was calculated in accordance with the standard scoring procedure, yielding the mean SUS score and Standard Deviation (SD) across all participants as described in [33]. For each participant, scores on the remaining questions, which were rated on a scale of 1 to 5, were calculated to assess the visual feedback component used for providing the BPB and the real-time monitoring of HR and BR (real-time monitoring was only accessed by the PTs). The mean of these scores was computed, taking into account all participants. These scores provided insights into the perceived impact and subjective quality of these different components.

5 Results

All ten patients and four PTs successfully rated the experimental PR sessions and completed the questionnaire. Table 3 summarizes the results categorized by area and statement. Overall, the enhanced system received positive feedback, attaining an average SUS score of 76.25±8.09 points for patients and 70.00±5.70 points for PTs.

While the scores from the questionnaire suggest a good degree of usability for the system evaluated, the qualitative data from free-statement transcripts[5] provided a greater context and support as to the reason that participants and PTs scored the system this way.

A. Overall System

(A.1) *I found it very useful.*
(A.2) *Very positive.*
(A.3) *It's a great experience and an asset for the future if I can use it.*
(A.4) *I think it would have been good to have tried the exercises beforehand to be familiar with them before evaluating specific components.*
(A.5) *It allows real-time monitoring during exercise and monitoring of activities that the patient performs independently.*
(A.6) *Education on synchronization of movements/breathing is important.*

[5] The statements from these participants were translated from Portuguese.

Table 3 Results on the questionnaire applied to the patients and to the PTs

Area and statement	Patients	PTs
System Usability Scale		
1. I think that I would like to use this system frequently.	4.1	4.5
2. I found the system unnecessarily complex.	1.6	1.75
3. I thought the system was easy to use.	4.4	3.75
4. I think that I would need help to be able to use this system.	3.1	3
I found the various functions in this system were well integrated.	4.3	2.75
6. I thought there was too much inconsistency in this system.	1.9	2.25
7. I would imagine that most people would learn to use this system very quickly.	3.7	3
8. I found the system very cumbersome to use.	1.2	1
9. I felt very confident using the system.	4.6	3.5
10. I needed to learn a lot of things before I could get going with this system.	3	1.75
Average total points	76.3/100	70/100
New information displayed		
A. I think that the quantity of the presented information is ideal.	4	4.25
B. Comparing the first session and the second session, I think that the visual element of the arrow changed my/patients' perception of how I/they should coordinate breathing and exercise.	3.3	3.5
C. Information about breathing rate and heart rate is useful for the patient during exercise.	–	2.5
D. Information about breathing rate and heart rate is useful for the physiotherapist during exercise.	–	4.75
E. I prefer that information about breathing rate and heart rate be available only on the physiotherapist's portal, which the patient does not have access to.	–	3.75
F. I could easily identify and understand breathing and heart rates.	–	3.5
Future perspectives		
G. I/The patient would be more motivated to coordinate breathing with exercise if there were a scoring system or rewards in the game.	3.9	4.5
H. It would be interesting to have a metric for breathing exercise coordination.	–	5
I. I can imagine myself/the patient using the system at home on my/his own, without the supervision of a physiotherapist.	3.2	4

Notes: The system under evaluation consists of the "breathing-version" of the game application, the used wearable, and real-time monitoring. The scoring is achieved using the Likert scale (1–5). 1-strongly disagree; 2-disagree; 3-neutral or no opinion; 4-agree; 5-strongly agree

B. Breathing Exercise Coordination

(B.1) *Difficult to control breathing with the exercises because it's the first contact.*

(B.2) *Sometimes I mix up the breathing timings.*

(B.3) *In some exercises, it's complicated to coordinate breathing with the exercises.*

C. Breathing Phase Biofeedback

(C.1) *The arrows help a lot in controlling breathing during exercises.*

(C.2) *It helped in coordinating breathing.*

(C.3) *Useful, although the symbolism of exhale-inhale is not obvious in its relevance to the exercise.*

(C.4) *I think the arrows, as they indicate direction, can confuse the patient between the respiratory moment and limb movement.*

(C.5) *The arrows help, but the transition from inhalation to exhalation is too fast.*

(C.6) *My attention was primarily focused on the balloons or the walls, so I think a more visually appealing representation accompanied by a sound signal could enhance the intended goal.*

(C.7) *I did not notice the arrow indicating breathing.*

(C.8) *People focus more on the game than on breathing.*

(C.9) *For some people, it can be motivating for the performance of respiratory therapy exercises, while others prefer to be verbally taught which exercises to do and don't give much importance to the system, thinking it complicates the execution.*

D. Suggestions and Refinements

(D.1) *Image resembling a lung.*

(D.2) *The use of a more visually appealing representation paired with sound feedback may enhance the intended outcome.*

(D.3) *Use a short or long vertical bar to indicate exhaling or inhaling.*

(D.4) *I think a non-directional indicator would be more advantageous (e.g., a balloon that inflates and deflates, a head with an arrow entering and exiting the mouth, or an auditory indication)*

(D.5) *A more noticeable arrow or something that draws attention to breathing.*

(D.6) *Short or long vertical bar (exhale or inhale).*

(D.7) *The change to be more paced between inhalation and exhalation.*

E. Monitoring

(E.1) *I didn't feel restricted.*

(E.2) *Any monitoring is essential.*

Moreover, in the PTs' quantitative evaluation of the patient's ability to coordinate breathing and exercise, an improvement was observed in 14 out of the 40 (i.e., 35%) exercise activities performed by the patients. Additionally, there was an overall enhancement in the BEC mean score from the first session (3.65/5) to the second session (4.23/5). Detailed results of this evaluation are presented in Table 4.

Table 4 PTs assessed the patient's ability to coordinate breathing and exercise in sessions 1 and 2, both without and with the use of BPB

	1 session				2 session			
Exercises	SF	SA	EF	LA	SF	SA	EF	LA
P1	5	5	2	5	5	5	5	5
P2	5	5	2	2	5	5	5	1
P3	5	5	1	1	5	5	5	5
P4	5	5	2	1	5	4	5	2
P5	5	4	5	2	5	5	5	5
P6	5	4	2	2	4	5	5	2
P7	5	5	4	2	5	5	5	2
P8	2	5	2	5	5	5	2	2
P9	5	5	5	2	5	5	2	2
P10	5	5	2	2	5	5	4	2
Mean	4.7	4.8	2.7	2.4	4.9	4.9	4.3	2.8
	Mean: 3.65/5				Mean: 4.23/5			

6 Discussion

This study investigated the perceptions of patients and PTs regarding the use of a custom-built interactive exercise biofeedback system integrated with wearable technology and implemented during PR. This research contributes to the existing body of literature, as only a limited number of studies have explored the opinions of end users, particularly in this population, and the utilization of physiological metrics (i.e., breathing phase, breathing, and heart rate), beyond motion and position. This is noteworthy, given the increasing use of sensors, wearables, and technological advancements. Furthermore, the results offer user-centered evidence that facilitates the optimization of engagement and effectiveness. By proactively identifying and addressing potential shortcomings, the study aims to enhance the overall understanding of the system's impact.

Patients and PTs generally exhibited a positive response to the use of such an enhanced system, reflected in good SUS scores. The system was well-received with enthusiasm. The results provide valuable insights that can inform the design of similar systems, offering suggestions to enhance the user experience and optimize the impact of including cardiorespiratory biofeedback.

Previous research has highlighted the importance of addressing various concerns when establishing a PTele solution. These requirements encompass adherence to PR core principles, adherence to quality and security standards, technological functionality, appropriateness of the environment, and technology literacy [34]. However, the results from this study suggest that a system built with a user-centered focus and considering the opinions of specialized healthcare professionals may be able to mitigate these concerns.

In relation to the new information presented (i.e., CRB), particularly the BPB, opinions were varied. Concerning the perception of improvement in BEC with the inclusion of the visual element (the arrow), the mean score was 3.3/5 for patients and 3.5/5 for PTs. This suggests that there is no clear consensus on this matter.

During the exercises, some patients encountered challenges in BEC, as revealed in statements B.1–B.3, such as *"In some exercises, it's complicated to coordinate breathing with the exercises"*. This further justifies the inclusion of cues and components to enhance and simplify the education and exercises for coordinating breathing and physical motion. Over the different experimental sessions, the PTs observed an improvement in this coordination in 14 out of 40 different exercises realized by the different patients (i.e., 10 participants over 4 different exercises) in the second session. Additionally, there was an overall improvement in the performance of all exercises from the first to the second session, with the most significant enhancements observed in Elbow Flexion (EF) and Leg Abduction (LA), compared to the improvements in Shoulder Flexion (SF) and Shoulder Abduction (SA).

Nevertheless, concerning the use of the upward and downward arrow as the component for breathing phase biofeedback, some patients acknowledge its benefits and utility, while others emphasize the potential for improvement and refinement (statements C.1–C.9). For example, some patient statements regarding the visual cues, such as *"The arrows help a lot in controlling breathing during exercises"* and *"It helped in coordinating breathing"*, indicate that participants found the feedback to be beneficial. However, other statements like *"Useful, although the symbolism of exhale-inhale is not obvious in its relevance to the exercise"* and *"I think the arrows, as they indicate a direction, can confuse the patient between the respiratory moment and limb movement"* revealed that some patients found it challenging and sometimes confusing.

Notably, patients generally encountered more difficulty in the BEC for exercises like EF and LA, as indicated by the session one mean scores in Table 4 (2.7 and 2.4, respectively). Although the BEC improved for EF and LA (from 2.7 to 4.3 and from 2.4 to 2.8, respectively), in these exercises, the used breathing cues—an upward and downward arrow (as described in Table 2)—displayed contradictory information to the exercise motion, which could be less intuitive. For instance, in those cases, a downward arrow indicated exhaling, while the exercise motion involved an upward motion. Therefore, it is suggested that an alternative breathing cue should be used to intuitively guide BEC, which can further enhance and facilitate this task.

Furthermore, some participants expressed that *"The arrows help, but the transition from inhalation to exhalation is too fast," "My attention was primarily focused on the balloons or the walls, so I think a more visually appealing representation accompanied by a sound signal could enhance the intended goal," "I did not notice the arrow indicating breathing,"* and *"People focus more on the game than on breathing."* These comments highlight the need to better integrate this feature into the specific game exercises, ensuring that it does not overwhelm users with this additional stimuli. These changes, in essence, are supported by Brennan et al. [14],

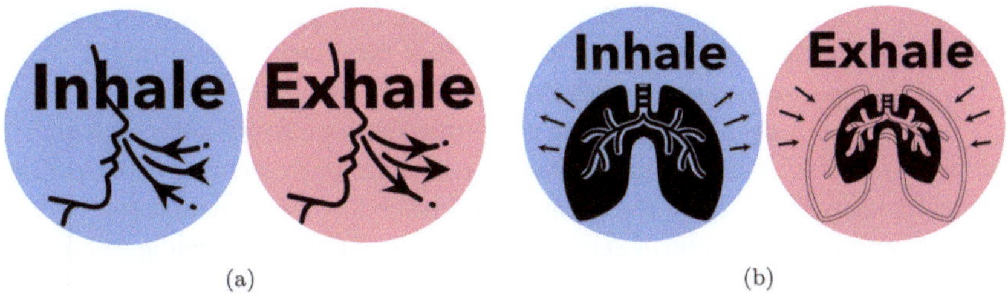

Fig. 5 Illustration of the suggested breathing biofeedback visual cues: (a) breathing lung visual cue; (b) breathing nose visual cue

which reports that the game should primarily provide feedback on the most critical aspects of the task.

Among 6 out of 14 patients and PTs, there is a belief that incorporating another element can offer better information for BEC. These insights provide constructive suggestions for alternative feedback cues or improvements, such as using an image resembling a lung, *"the suggestion of using an element that doesn't rely on directional cues, like a balloon that inflates and deflates, a head with an arrow entering and exiting the mouth, or auditory cues," "the use of a more visually appealing representation paired with sound feedback may enhance the intended outcome," "use a short or long vertical bar to indicate exhaling or inhaling"* (as illustrated in Fig. 5). It is worth exploring the integration of the biofeedback cue as a game element or an image that conveys better the perception of the task, potentially inflating and deflating a lung, along with audio feedback, as this may provide a more closely aligned feedback experience to what is provided by PTs in a clinical setting [35, 36].

Regarding the monitoring with the chest wearable, it was well-received, with a patient even stating, *"I didn't feel restricted"*; *"any monitoring is essential"*, commenting on the comfort and nonintrusiveness of the wearable device.

The "future perspectives" questionnaire category indicated a shared interest among both patients and PTs in incorporating a gamified reward component and developing a system to evaluate performance based on movement during breathing phases. The integration of this feature will involve combining information from the provided breathing guidance (further described in Sect. 4.2.2) with the monitored breathing phase (derived from the wearable Sect. 4.2.1) while guaranteeing synchronous data acquisition.

Despite the experimental efforts to divide the experimental session, into two parts, aimed at clarifying the intricacies of the new implementation related to CRB, the study conducted has certain limitations that must be acknowledged when drawing conclusions. First, the cardiorespiratory monitoring, as it was not fully integrated into the Clynx platform (it was displayed on a second screen), can behave as an additional distraction for the PTs. Second, participants had no previous contact with this telerehabilitation system.

Ensuring that users are well-acquainted and familiarized with the system may be crucial in such studies because, without this familiarity, users might perceive the new implementation as overwhelming making it challenging to isolate and assess its specific impact. Recognizing this limitation emphasizes the importance of considering user familiarity in future studies to obtain more accurate and meaningful conclusions regarding the implementation of biofeedback enhancements.

Notwithstanding the heterogeneous pathological COPD of the participants, it would be of value to include more participants to attain parity and to gain further insights and opinions on the system.

Moreover, this proof-of-concept lays the foundation for future longitudinal studies aimed at analyzing potential clinical outcomes, which compare the innovative rehabilitation approach with standard care, as discussed by other authors [7, 37]. Clarifying its clinical value can promote its generalized adoption in the rehabilitation context [38]. Additionally, future research could explore the feasibility of autonomous use by patients (after familiarization) outside the hospital environment, such as in their homes, which, ultimately, could enhance the efficacy and durability of exercise training improvements and also facilitate the accessibility of those who need care. Ultimately, outside the cohort of COPD patients, telerehabilitation enhanced with cardiorespiratory biofeedback could be used by patients with other CRDs prescribed for PR.

7 Summary

The rise of technology and wearable sensors presents an opportunity to introduce innovative and effective ways of supporting patients in PR exercise training. In this pilot study, the implementation of cardiorespiratory biofeedback in a gamified telerehabilitation system was perceived as valuable by both patients and physiotherapists. This addition offered an engaging and motivating solution that could significantly influence patient PR experience. Drawing on input from intervenients, patients, and physiotherapists, this study focused on a user-centered experience. This approach revealed potential refinements for the future, enhancing and amplifying the overall user experience. Notably, insights from users shed light on possible shortcomings, particularly in the used biofeedback component, enabling early identification and improvement in the development cycle.

Furthermore, establishing the clinical value and demonstrating the real-world accuracy of the telerehabilitation system, enhanced with cardiorespiratory biofeedback will provide crucial evidence. This evidence, once clarified, has the potential to drive widespread adoption of such systems in PR, contributing to its broader and more effective use and improving patients' accessibility to PR. The main takeaways are as follows:

- Project objective: This project aims to enhance and adapt an existing telerehabilitation solution for pulmonary rehabilitation by providing continuous real-time monitoring of vital signs, using wearable technology, and feedback on breathing exercise coordination.

- System components: This system includes the Clynx Platform, which offers a set of gamified musculoskeletal exercises and uses a depth sensor camera, real-time heart and breathing rate monitoring using a Polar soft chest strap paired with a CardioScientISST unit and a previously validated breathing sensor, and implementation of breathing exercise coordination (BEC) guidelines.
- Technical implementation: The ScientISST API and Bluetooth connection enabled real-time heart and breathing rate computation. Heart rate was computed based on the BioSPPy toolbox Hamilton segmenter implementation. Breathing rate computation used the Automatic Multiscale-Based Peak Detection (AMPD) algorithm. Breathing exercise coordination (BEC) guidance, implemented a breathing phase biofeedback, was designed with clinician expertise and featured graphical cues—an up arrow for inhalation and a down arrow for exhalation—automatically triggered by maximum joint angles detected via the depth camera.
- Pilot testing: Ten COPD patients were recruited to test the system in a real pulmonary rehabilitation (PR) setting under clinician supervision. This allowed assessment of feasibility, usability, and impact and provided feedback for improvement. Testing involved two sessions, each with four PR exercises (three upper body and one leg exercise), with one session without BEC guidance and the other including BEC guidance, but both with BR and HR monitoring. For each patient, case report files and quantitative and qualitative questionnaires were completed, with additional open comments and statements recorded. Clinicians were also asked to complete similar documentation reflecting their experiences.
- Feedback and results: The enhanced system received positive feedback, achieving an average SUS score of 76.25 ± 8.09 points for patients and 70.00 ± 5.70 points for physical therapists. According to clinicians' qualitative feedback, coordinated breathing and exercise improved in 35% (14 out of 40) of the activities performed by patients. Additionally, the BEC mean score increased from the first session (3.65/5) to the second session (4.23/5).
- Limitations and recommendations: Some limitations and recommendations emerged from this study. There was no consensus on the specific breathing phase guidance element—feedback elements should be clear and easily understood by users. It is also essential to ensure that feedback in the rehabilitation game focuses on the most critical aspects of the task, potentially including gamification specifically designed for breathing phase coordination. Familiarization with the non-enhanced rehabilitation system was identified as beneficial for a clearer understanding of the enhancements and adaptations studied.
- Overall and future impact: The overall system and the concept of system adaptation were perceived as valuable in the context of pulmonary rehabilitation. Demonstrating its usability and clinical value in a telerehabilitation setting, where the system components are seamlessly integrated, may encourage broader adoption.

Author Contributions Inês Carvalho and Fátima Rodrigues conceived and designed the study; Ana Martins and Rute Rodrigues conducted the recruitment; Inês Carvalho, Ana Martins, and Rute Rodrigues conducted data collection; Inês

Carvalho contributed to the development and implementation of the biofeedback with the assistance of Gonçalo Chambel; Ana Sofia developed the prototype for physiological monitoring; Inês Carvalho conducted the data analyses and drafted the manuscript; Ana Sofia Cacais do Carmo, Mariana Abreu, and Patrícia Bota reviewed and adapted the manuscript; Fátima Rodrigues, Hugo Silva, and Joana Pinto supervised the conducted research.

Acknowledgments This work was partially funded by FCT/MECI through national funds (FCT grants 2020.06675.B and 2021.08297.BD, and by FCT project reference and DOI identifier https://doi.org/10.54499/2022.12369.BD) and when applicable co-funded EU funds under project UID/50008: Instituto de Telecomunicações. It was also co-supported by the European Regional Development Fund (FEDER), through the Lisbon Regional Programme (LISBOA 2030) of the Portugal 2030 framework - LISBOA2030-FEDER-01318700 (ComSense).

References

1. M. Fekete, V. Fazekas-Pongor, P. Balazs, S. Tarantini, A.N. Nemeth, J.T. Varga, Role of new digital technologies and telemedicine in pulmonary rehabilitation: smart devices in the treatment of chronic respiratory diseases. Wiener Klinische Wochenschrift **133**, 1201–1207 (2021)
2. T. Bonnevie, P. Smondack, M. Elkins, B. Gouel, C. Medrinal, Y. Combret, J.F. Muir, A. Cuvelier, G. Prieur, F.E. Gravier, Advanced telehealth technology improves home-based exercise therapy for people with stable chronic obstructive pulmonary disease: a systematic review. J. Physiother. **67**, 27–40 (2021)
3. A. Rossi, B. Butorac-Petanjek, M. Chilosi, B.G. Cosío, M. Flezar, N. Koulouris, J. Marin, N. Miculinic, G. Polese, M. Samaržija, S. Skrgat, T. Vassilakopoulos, A. Vukić-Dugac, S. Zakynthinos, M. Miravitlles, Chronic obstructive pulmonary disease with mild airflow limitation: current knowledge and proposal for future research—a consensus document from six scientific societies. Int. J. COPD **12**, 2593–2610 (2017)
4. B. Mccarthy, D. Casey, D. Devane, K. Murphy, E. Murphy, Y. Lacasse, Pulmonary rehabilitation for chronic obstructive pulmonary disease, in *Cochrane Database of Systematic Reviews*, vol. 2015 (2015)
5. A. Dimitrova, N. Izov, I. Maznev, D. Vasileva, M. Nikolova, Physiotherapy in patients with chronic obstructive pulmonary disease. Open Access Macedonian J. Med. Sci. **5**, 720–723 (2017)
6. Martijn Spruit and Linda Nici, *Textbook of Pulmonary Rehabilitation*, ch. Current Concepts and Definitions (Springer, Berlin, 2018), pp. 19–22
7. N.S. Cox, S.D. Corso, H. Hansen, C.F. McDonald, C.J. Hill, P. Zanaboni, J.A. Alison, P. O'Halloran, H. Macdonald, A.E. Holland, Telerehabilitation for chronic respiratory disease. Cochrane Database System. Rev. **2021**, CD013040 (2021)
8. S. Rutkowski, Management challenges in chronic obstructive pulmonary disease in the COVID-19 pandemic: telehealth and virtual reality. J. Clin. Med. **10**, 1–11 (2021)
9. C.-K. Tey, J. An, W.-Y. Chung, A novel remote rehabilitation system with the fusion of noninvasive wearable device and motion sensing for pulmonary patients. Comput. Math. Methods Med. **2017**, 5823740 (2017)
10. C.-H. Shih, N. Tomita, Y.X. Lukic, Á. H. Reguera, E. Fleisch, T. Kowatsch, Breeze: smartphone-based acoustic real-time detection of breathing phases for a gamified biofeedback breathing training, in *Proceedings. of the ACM on Interactive, Mobile, Wearable and Ubiquitous Technologies*, vol. 3(4) (2019), pp. 1–30

11. V. Colombo, A. Aliverti, M. Sacco, Virtual reality for COPD rehabilitation: a technological perspective. Pulmonology **28**(2), 119–133 (2022)
12. D. Calvaresi, M. Marinoni, A.F. Dragoni, R. Hilfiker, M. Schumacher, Real-time multi-agent systems for telerehabilitation scenarios. Artif. Intell. Med. **96**, 217–231 (2019)
13. N. Niknejad, W.B. Ismail, A. Mardani, H. Liao, I. Ghani, A comprehensive overview of smart wearables: the state of the art literature, recent advances, and future challenges. Eng. Appl. Artif. Intell. **90**, 103529 (2020)
14. L. Brennan, E.D. Zubiete, B. Caulfield, Feedback design in targeted exercise digital biofeedback systems for home rehabilitation: a scoping review. Sensors **20**, 181 (2020)
15. S.-C. Chen, R.-J. Liing, K. Lin, T.-W. Lu, Immediate effects of arm abduction synchronized with breathing movement on the chest wall kinematics in healthy adults. J. Med. Biol. Eng. **32**, 309–312 (2012)
16. T.M. Annaswamy, G.N. Pradhan, K. Chakka, N. Khargonkar, A. Borresen, B. Prabhakaran, Using biometric technology for telehealth and telerehabilitation. Phys. Med. Rehab. Clin. North Am. **32**, 437–449 (2021)
17. K. Kulkarni, J.H. Nichols, A.A. Armoundas, J.D. Roberts, RespiCo: a novel, flexible, and stand-alone electronic respiratory coaching device. HardwareX **12**, e00335 (2022)
18. A. Khamis, B. Kusy, C.T. Chou, W. Hu, WiRelax: towards real-time respiratory biofeedback during meditation using WiFi. Ad Hoc Netw. **107**(2), 102226 (2020)
19. D.K. Arvind, T. Georgescu, C.A. Bates, D. Fischer, Q. Zhou, Home-based pulmonary rehabilitation of COPD individuals using the wearable respeck monitor, in *Body Area Networks. Smart IoT and Big Data for Intelligent Health Management*, ed. by M. Ur Rehman, A. Zoha (Springer, Berlin, 2022), pp. 176–191
20. C.J. Hill, M. Lazzeri, F. D'Abrosca, *Textbook of Pulmonary Rehabilitation*, ch. Breathing Exercises and Mucus Clearance Techniques in Pulmonary Rehabilitation (Springer, Berlin, 2018), pp. 205–216
21. World Medical Association, World Medical Association Declaration of Helsinki: ethical principles for medical research involving human subjects. JAMA **310**, 2191–2194 (2013)
22. GOLD, Global initiative for chronic obstructive lung disease global strategy for the diagnosis, management, and prevention of chronic obstructive pulmonary disease. Tech. rep., Global Initiative for Chronic Obstructive Lung Disease, 2023
23. L. Pereira, F. de Melo, F.A. Santos, A.F. Ferreira, H.P. da Silva, Scientisst core: a novel hardware development platform for biomedical engineering. HardwareX **21**, e00630 (2025)
24. A.S. Carmo, I. Carvalho, J.F. Pinto, G. Chambel, A. Martins, R. Roberto, H. Ferreira, F. Rodrigues, A. Fred, H.P. da Silva, A magnetic field-based wearable respiration sensor for real-time monitoring during pulmonary rehabilitation. IEEE Trans. Biomed. Eng. **71**(7), 2243–2252 (2024)
25. P. Hamilton, Open source ECG analysis, in *Computers in Cardiology* (2002), pp. 101–104
26. P. Bota, R. Silva, C. Carreiras, A. Fred, H.P. da Silva, Biosppy: a python toolbox for physiological signal processing. SoftwareX **26**, 101712 (2024)
27. F. Scholkmann, J. Boss, M. Wolf, An efficient algorithm for automatic peak detection in noisy periodic and quasi-periodic signals. Algorithms **5**(4), 588–603 (2012)
28. C. Garvey, M.P. Bayles, L.F. Hamm, K. Hill, A. Holland, T.M. Limberg, M.A. Spruit, Pulmonary rehabilitation exercise prescription in chronic obstructive pulmonary disease. J. Cardiopulmonary Rehab. Prevention **36**, 75–83 (2016)
29. K.R. Kendrick, S.C. Baxi, R.M. Smith, Usefulness of the modified 0-10 Borg scale in assessing the degree of dyspnea in patients with COPD and asthma. J. Emer. Nursing **26**(3), 216–222 (2000)
30. G.M. Sullivan, A.R. Artino, Analyzing and interpreting data from likert-type scales. J. Graduate Med. Educ. **5**(4), 541–542 (2013)
31. R. Argent, P. Slevin, A. Bevilacqua, M. Neligan, A. Daly, B. Caulfield, Wearable sensor-based exercise biofeedback for orthopaedic rehabilitation: a mixed methods user evaluation of a prototype system. Sensors **19**(2), 432 (2019)

32. L. Liu, X. Chen, Z. Lu, S. Cao, D. Wu, X. Zhang, Development of an EMG-ACC-based upper limb rehabilitation training system. IEEE Trans. Neural Syst. Rehab. Eng. **25**, 244–253 (2017)
33. J. Brooke, SUS: a quick and dirty'usability. Usability Eval. Ind. **189**(3), 189–194 (1996)
34. C.D. Santos, F. Rodrigues, C. Caneiras, C. Bárbara, From inception to implementation: Strategies for setting up pulmonary telerehabilitation. Front. Rehab. Sci. **3**, 2673–6861 (2022)
35. A. Nicolò, C. Massaroni, E. Schena, M. Sacchetti, The importance of respiratory rate monitoring: from healthcare to sport and exercise. Sensors **20**, 1–45 (2020)
36. G. Spina, G. Huang, A. Vaes, M. Spruit, O. Amft, COPDTrainer: a smartphone-based motion rehabilitation training system with real-time acoustic feedback, in *Proceedings of the ACM Int. Joint Conference on Pervasive and Ubiquitous Computing* (2013), pp. 597–606
37. I. Carpinella, D. Cattaneo, G. Bonora, T. Bowman, L. Martina, A. Montesano, M. Ferrarin, Wearable sensor-based biofeedback training for balance and gait in parkinson disease: a pilot randomized controlled trial. Arch. Phys. Med. Rehab.**98**, 622–630, (2017)
38. F. Michard, A sneak peek into digital innovations and wearable sensors for cardiac monitoring. Clin. Monitor. Comput. **31**, 253–259 (2017)

Index

H. P. da Silva et al. (eds.), *Open Source Biomedical Engineering*,
https://doi.org/10.1007/978-3-032-03655-1

MIX
Papier aus verantwortungsvollen Quellen
Paper from responsible sources
FSC® C105338

If you have any concerns about our products,
you can contact us on
ProductSafety@springernature.com

In case Publisher is established outside the EU,
the EU authorized representative is:
Springer Nature Customer Service Center GmbH
Europaplatz 3, 69115 Heidelberg, Germany

Printed by Libri Plureos GmbH
in Hamburg, Germany